SECOND
EDITION

Knowledge Management Handbook

Collaboration and Social Networking

T0172975

SECOND
EDITION

Knowledge Management Handbook

Collaboration and Social Networking

Edited by
Jay Liebowitz

 CRC Press
Taylor & Francis Group
Boca Raton London New York

CRC Press is an imprint of the
Taylor & Francis Group, an **informa** business

CRC Press
Taylor & Francis Group
6000 Broken Sound Parkway NW, Suite 300
Boca Raton, FL 33487-2742

First issued in paperback 2019

© 2012 by Taylor & Francis Group, LLC
CRC Press is an imprint of Taylor & Francis Group, an Informa business

No claim to original U.S. Government works

ISBN-13: 978-1-4398-7803-3 (hbk)
ISBN-13: 978-0-367-38122-6 (pbk)

Library of Congress Cataloging-in-Publication Data

Knowledge management handbook : collaboration and social networking / editor, Jay
  Liebowitz. -- 2nd ed.
       p. cm.
  Includes bibliographical references and index.
  ISBN 978-1-4398-7803-3 (hardcover : alk. paper)
  1. Knowledge management. I. Liebowitz, Jay, 1957-

HD30.2.K637 2012
658.4'038--dc23   2012014561

Visit the Taylor & Francis Web site at
http://www.taylorandfrancis.com

and the CRC Press Web site at
http://www.crcpress.com

Contents

Preface

So, after having the first-ever *Knowledge Management Handbook* published back in 1999, why is there now, after these long awaited years, a second edition that focuses on collaboration and social networking? The question about having a second edition of the *Knowledge Management Handbook* after so many years have passed certainly resonated in my own mind. Certainly, other excellent handbooks on knowledge management (KM) have been published over the ensuing years. However, because organizations are gaining traction via KM through collaboration and social networking, I thought it may be worthwhile to assemble some leading cases and chapters worldwide that focus strictly on these two areas of greatest KM impact.

This edition of the *Knowledge Management Handbook* tries to accomplish this task to further complement the existing references and sources in the KM literature.

The research shows that collaboration and social networking foster knowledge sharing and innovation. Through the use of online communities of practice, for example, collaboration and knowledge sharing are activated to catalyze new connections, ideas, and practices. Social and organizational network analysis shows that the relationships established outside one's own area ("weak ties") are where the innovation often occurs. Through knowledge management, the hope is that further collaboration and social networking will result in increasing innovation.

Already, the recent literature points to innovation resulting from knowledge sharing initiatives. In Yang's [1] work on the effect of KM on product innovation, it was shown that internal knowledge sharing and external knowledge acquisition positively complement product innovation. Kianto's [2] study of 54 small and medium-sized organizations from various industries found that knowledge sharing and knowledge acquisition are enablers of continuous innovation. They also showed that knowledge sharing practices among colleagues are the most significant predictor of self-rated continuous innovation [2]. Cross et al. [3] showed that "energizing" employees helps to drive innovation. Wang and Wang [4] found that explicit knowledge sharing has more significant effects on innovation speed and performance, while tacit knowledge sharing has more significant effects on innovation quality and operational performance.

In organizations, as shown throughout this handbook, knowledge management initiatives produced positive results. For example, Tata Chemicals Ltd. in India [5] demonstrated the benefits of their KM efforts by linking KM to more innovation (increased number of patents), improved quality (more process conscious), improved problem solving (competency improvement), and improved productivity (from 2010 to 2012, expected savings from KM efforts should be $2.35 million). The average number of visits in their KM portal per month in 2010 to 2011 was 197,250.

Even though the research shows various benefits of using knowledge management, there are still skeptics of what it can offer. Certainly, more work is needed in developing outcome metrics to measure the value of these knowledge sharing efforts to the organization's strategies, goals, and objectives. Even when we look at collaboration and social networking, the Katzenbach Partners [6] (who joined Booz & Company) point out that the informal organization is often misunderstood and poorly managed. Thus, this handbook serves to provide further evidence and examples of how collaboration and social networking can be better understood for organizational gain, and also points out some areas the KM community needs to further address.

There are many people to thank for making this second edition of the *Knowledge Management Handbook* possible. First, without constant insistence from my publisher, Nora Konopka, with CRC Press, this second edition would never have been possible. Nora felt strongly that there was a home and need for this second edition, and I hope that we are able to serve the broader KM community by focusing on the themes of collaboration and social networking. I greatly appreciate the excellent work of the CRC Press publishing staff. Second, I need to thank the wonderful contributors of the chapters and cases in this handbook. They are among the leaders worldwide in their field, so we are very pleased to be able to highlight their work. Third, I thank my wife, Janet, and my family for their great encouragement and support. And finally, I would like to thank my students and colleagues at the University of Maryland University College, as well as my outside professional contacts, for helping me provide a solid foundation for encouraging research and scholarship.

Well, as I told our sons, Jason and Kenny, growing up, "Have fun, learn a lot, get an A!" We hope you enjoy this volume!

Jay Liebowitz, D.Sc.
Washington, DC

REFERENCES

1. Yang, D. (2011), The effect of knowledge management on product innovation—Evidence from the Chinese software outsourcing vendors, *iBusiness Journal*, Vol. 3, Scientific Research.
2. Kianto, A. (2011), The influence of knowledge management on continuous innovation, *International Journal of Technology Management*, Vol. 55, No. 1/2.
3. Cross, R., J. Linder, and A. Parker (2006), Charged Up Managing the Energy That Drives Innovation, Network Roundtable White Paper, University of Virginia, Charlottesville.
4. Wang, Z., and N. Wang (2011), Knowledge sharing, innovation, and firm performance, Submitted to the *Expert Systems with Applications: An International Journal*.
5. Kruthiventi, D., and B. Sudhakar (2011), "Insight to Intelligence to Innovation—KM Journey at TCL," APQC Knowledge Management 2011 Conference, Houston, TX.
6. Katzenbach, J., and Z. Khan (2009), *Leading Outside the Lines: How to Mobilize the (in)Formal Organization*, Jossey-Bass, Hoboken, NJ.

The Editor

Jay Liebowitz is the Orkand Endowed Chair of Management and Technology in the Graduate School of Management and Technology at the University of Maryland University College (UMUC). He previously served as a professor in the Carey Business School at Johns Hopkins University, Baltimore, MD. He is ranked one of the top 10 knowledge management (KM) researchers/practitioners out of 11,000 worldwide and was ranked number two in KM strategy worldwide according to the January 2010 *Journal of Knowledge Management*. At Johns Hopkins University, he was the founding Program Director for the Graduate Certificate in Competitive Intelligence and the Capstone Director of the MS-Information and Telecommunications Systems for Business Program, where he engaged over 30 organizations in industry, government, and not-for-profits in capstone projects.

Prior to joining Johns Hopkins, Liebowitz was the first Knowledge Management Officer at the National Aeronautics and Space Administration (NASA) Goddard Space Flight Center. Before NASA, Liebowitz was the Robert W. Deutsch Distinguished Professor of Information Systems at the University of Maryland–Baltimore County, Professor of Management Science at George Washington University, and Chair of Artificial Intelligence at the U.S. Army War College.

Liebowitz is the founder and editor-in-chief of *Expert Systems with Applications: An International Journal* (published by Elsevier), which is ranked third worldwide for intelligent systems/artificial intelligence (AI)-related journals, according to the most recent Thomson impact factors. He is a Fulbright Scholar, IEEE-USA Federal Communications Commission Executive Fellow, and Computer Educator of the Year (International Association for Computer Information Systems). He published over 40 books and a myriad of journal articles on knowledge management, intelligent systems, and information technology (IT) management. His most recent books are *Knowledge Retention: Strategies and Solutions* (Taylor and Francis, 2009), *Knowledge Management in Public Health* (Taylor and Francis, 2010), *Knowledge Management and E-Learning* (Taylor and Francis, 2011), and *Beyond Knowledge*

Management: What Every Leader Should Know (Taylor and Francis, 2012). In October 2011, the International Association for Computer Information Systems (IACIS) named the "Jay Liebowitz Outstanding Student Research Award" for the best student research paper at the IACIS annual conference. He has lectured and consulted worldwide.

Contributors

Wolfgang Aigner
Institute of Software Technology
and Interactive Systems
Technical University
Vienna, Austria

Michael Barrett
Judge Business School
University of Cambridge
Cambridge, United Kingdom

Ulrike Becker-Kornstaedt
Fraunhofer Center for
Experimental Software
Engineering
College Park, Maryland

Denise A.D. Bedford
Kent State University
Kent, Ohio

Melvin Brown II
MC Brown and Associates, LLC
Manassas, Virginia

Andrew Campbell
Applied Knowledge Group, Inc.
Reston, Virginia

Francisco J. Cantú
Tecnologico de Monterrey
Monterrey, Nuevo León, Mexico

Héctor G. Ceballos
Tecnologico de Monterrey
Monterrey, Nuevo León, Mexico

Katherine Clark
Virginia Department of
Transportation
Charlottesville, Virginia

Kimiz Dalkir
McGill University
Graduate School of Library and
Information Studies
Montreal, Quebec, Canada

Paolo Federico
Institute of Software Technology
and Interactive Systems
Technical University
Vienna, Austria

Maureen Hammer
Virginia Department of
Transportation
Charlottesville, Virginia

Bradley Hilton
U.S. Army
Lansing, Kansas

Louise Humphreys
Price Modern LLC
Baltimore, Maryland

Molly Jackson
Community Analytics
Baltimore, Maryland

Devsen Kruthiventi
Tata Chemicals Ltd.
Andhra Pradesh, India

Moria Levy
ROM Knowledgeware Ltd.
Reut, Israel

Jay Liebowitz
Graduate School of Management
    and Technology
University of Maryland University
    College
Adelphi, Maryland

Celine Miani
Judge Business School
University of Cambridge
Cambridge, United Kingdom

Silvia Miksch
Institute of Software Technology
    and Interactive Systems
Technical University
Vienna, Austria

Myra Norton
Community Analytics
Baltimore, Maryland

Theresa C. Norton
Jhpiego (an affiliate of Johns
    Hopkins University)
Baltimore, Maryland

Eivor Oborn
School of Management
Royal Holloway
University of London
Egham, United Kingdom

Jürgen Pfeffer
CASOS Center, Institute for
    Software Research
Carnegie Mellon University
Pittsburgh, Pennsylvania

Gloria Phillips-Wren
Information Systems and
    Operations Management
Sellinger School of Business and
    Management
Loyola University Maryland
Baltimore, Maryland

Michael Prevou
Strategic Knowledge Solutions
Leavenworth, Kansas

Forrest Shull
Fraunhofer Center for Experimental
    Software Engineering
College Park, Maryland

Michael Smuc
Department of Knowledge and
    Communication Management
Danube University
Krems, Austria

B. Sudhakar
Tata Chemicals Ltd.
Bombay House
Mumbai, India

Florian Windhager
Department of Knowledge
and Communication
Management
Danube University
Krems, Austria

Doug Wise
Community Analytics
Baltimore, Maryland

Markos Zachariadis
Judge Business School
University of Cambridge
Cambridge, United Kingdom

Lukas Zenk
Department of Knowledge and
Communication Management
Danube University
Krems, Austria

1

Collaboration and Social Networking: The Keys to Knowledge Management— Introductory Thoughts

Jay Liebowitz

INTRODUCTION

The heart of knowledge management (KM) is sharing knowledge, making connections, and generating new ideas through collaboration and interaction. Along these lines, there are two schools of thought: the codification "school" and the "personalization" school. The codification camp emphasizes systems-oriented approaches to knowledge management that focus on "collection." The personalization camp focuses on the people-to-people "connection" approach. Ultimately, organizations should apply both approaches to knowledge management, but typically one of the two will take dominance.

In this chapter, and throughout the *Knowledge Management Handbook*, we are advocating the personalization "connection" approach—collaboration and social networking. At a conference on Social Media Strategy at George Washington University in the Spring 2011 semester, the experts there felt

- Within three years, more people will surf the Web on mobile devices than on computers
- For people under the age of 40, nearly half of all media consumption takes place online

Forrester (www.forrester.com) predicts that social networking will account for nearly half of the $4.6 billion market it forecasts for Web 2.0 products by 2013.

The social media craze is already upon us, as further witnessed by the initial public offering (IPO) in May 2011 of LinkedIn, soon to be followed by potential IPOs of other social media sites. Mark Zuckerberg, the chief executive officer (CEO) of Facebook, was the 2010 *Time* Person of the Year. Social networking and collaboration are the lingua franca of today's world, especially among the Gen Y'ers. The Horizon Report (http://wp.nmc.org/horizon2011/sections/executive-summary/) tracks learning and knowledge, and they are crossing geographical borders through online collaborative workspaces and social networking. Blogs, wikis, and Twitter are all part of the milieu of further connecting and collaborating with others. And certainly, mobile computing will continue to play an important role in how we operate, both at work and home.

THE INFORMAL ORGANIZATION

Your people are your competitive edge. This has been echoed time and again by such CEOs as Jim Goodnight, the CEO of SAS, who said "about 95% of my assets walk out the front door every evening and my job is to bring them back the next day" (F. Leistner, 2010, *Mastering Organizational Knowledge Flow: How to Make Knowledge Sharing Work*, Wiley, New York). Gartner (September 2010) also emphasizes the importance of human capital as stated by views on knowledge management:

- As KM program leaders determine their goals, they should consider how the emphasis has shifted away from just collecting content and refocus on connecting people with the content or people they need.
- KM leaders need to shift their focus to the management and people-related issues that contribute to successful implementation.

One might think that organizations would be investing resources to capitalize and catalyze connections and collaborations for further strengthening the informal organization. However, this may not be the case. According to the American Society for Training and Development (ASTD) and the Institute for Corporate Productivity (www.i4cp.com), they found in August 2009 in their research on informal learning that 98% of those surveyed say that informal learning enhances employee performance. However, 36% dedicated no money to informal learning and 78% dedicated 10% or less of the training budget to it. There was also a Web 2.0 study by ASTD, i4cp, and Booz Allen which found only a small minority of companies are using Web 2.0 technologies in learning.

In the recent book by the Katzenbach Partners titled *Leading Outside the Lines: How to Mobilize the Informal Organization, Energize Your Team, and Get Better Results* (John Wiley, 2010), the authors found that the informal organization is poorly understood and managed in most corporate settings. Five signs that your informal organization is alive and well, according to the Katzenbach Partners, are

- The word gets out fast.
- "Change" is not a dirty word.
- Collaboration is the default mode.
- Employees are tapped in.
- Stories demonstrate values.

Organizations need to pay close attention to nourishing the informal networks as often the innovation takes place from the relationships developed from outside one's own discipline or department. In social network analysis parlance, we say that the "weak ties" are where the innovation often happens through collaboration from outside one's area. Collaboration across the organizational boundaries can promote new ideas through capitalizing on different perspectives and backgrounds.

SHORT CASE: APPLYING KNOWLEDGE SHARING AT MY UNIVERSITY

What is the "university model" for knowledge management and what should it be? Let me explain this through a recent case of mine dealing with how to foster university research and scholarship through knowledge sharing activities, thus building the informal networks among the faculty at the university. This was selected as a top finalist in the USAID KM Impact Challenge (http://kdid.org/kmic).

In my role as the Orkand Chair in Management and Technology, I was asked to spearhead and develop a strategy to further promote research and scholarship at my university, the University of Maryland University College (UMUC). With my background in knowledge management, this was a perfect opportunity to build and nurture a knowledge sharing culture for research and scholarship in order to complement our core teaching mission. UMUC has about 94,000 students and 3,000 faculty members in 28 countries. Most of our courses are through e-learning. The KM initiative thus unfolded to be creating a knowledge sharing environment for promoting research and scholarship activities primarily among our full-time faculty and our doctoral students. In order to make this a reality, the following implementation components, among others, were applied to focus on internal awareness and external perspectives:

- Created a one-stop access UMUC research and scholarship site, through the help of the library: http://www.umuc.edu/library/research_pubs/research.shtml
- Created Flash objects of three KM-related talks, through the help of our Multimedia Faculty Lab, in order to spread further awareness of KM and insert these concepts throughout our appropriate undergraduate and graduate-level curricula: http://polaris.umuc.edu/de/csi/2010_JayLiebowitz/ppt_syn/JayLiebowitz_index3.html
- Organized monthly brown bag "Lunch and Learn" faculty research seminars, visiting scholar lectures, Orkand Chair distinguished lecture series, UMUC working papers series, Provost's best paper competition, and a faculty research grant program.
- Helped in creating SOARS (Student Opportunities to Advance Research and Scholarship), whereby our doctoral students present their work each semester in a poster session on campus.
- Created the first annual ShareFair (face-to-face and virtual) on UMUC Faculty Research and Scholarship to include faculty research poster sessions, knowledge cafes (moderated round table discussions), keynote address on cyberlearning trends and research directions, journal and book editors' panel on how to get published, and recognition awards for knowledge sharing and research and scholarship.

Besides anecdotal evidence of our faculty expressing delight in learning new things, generating new ideas through collaborating with others outside their own fields, and simply enjoying the interactions created by some of the above activities, we try to assess outcome measures for our knowledge sharing activities (aside from the system and output measures calculated by use of Google Analytics and through our annual UMUC Research and Scholarship Inventory Survey). A key outcome measure in a university with over 300 doctoral students should be "knowledge generation." Even though UMUC is more of a "teaching" university versus a "research" university, we still must have a cadre of research-active faculty in order to support our doctoral program (we have a scholar-practitioner-focused DM [Doctor of Management] program versus the traditional PhD program). Knowledge generation, in this context, would mean developing new ways of thinking, synthesizing, and generating new ideas. New metrics to make this happen are needed, such as

- The number of colleague-to-colleague relationships spawned through the various knowledge exchange/sharing activities previously mentioned (especially important would be the relationships developed of faculty outside one's own field for possibly increasing innovation)
- The reuse rate of "frequently accessed/reused" knowledge on research projects and in teaching resulting in efficiencies and effectiveness
- The capture of key expertise in an online way (i.e., the number of key concepts converted from tacit to explicit knowledge in the knowledge repositories and used by those in the university)
- The dissemination of knowledge sharing (i.e., distribution of knowledge) to appropriate individuals
- The number of new ideas generating innovative teaching, research, and departmental projects
- The number of lessons learned and best practices applied to create value added (i.e., decreased proposal writing/development time, increased student/faculty satisfaction, etc.)
- The number of "serious" anecdotes presented about the value of the organization's KM initiatives
- The number of "apprentices (students or junior faculty)" that one mentors, and the success of these apprentices as they mature in the university

These measures and others can serve to further formulate a "university model" for knowledge management. Certainly, many of these measures and knowledge sharing initiatives can be adapted to fit organizations such as think tanks, consulting firms, government agencies, not-for-profits, and companies. But, before basking in this "knowledge management light" for others, we need to do a better job in academe for espousing knowledge sharing tenets and trying creative measures for knowledge capture, sharing, application, and generation. We can combine the theory and practice of knowledge management as a test-bed, and, if successful, then apply these concepts to other organizations. If the next generation is to surpass the previous one, universities have a central role to play, and we can start by applying these concepts in-house so that a "university model" for knowledge management can be created and shared by others.

WHAT ARE SOME OTHERS DOING IN TERMS OF PROMOTING COLLABORATION AND KNOWLEDGE SHARING?

Various organizations are taking a proactive stance toward promoting collaboration and knowledge sharing. According to the report titled "Knowledge Management Performance Metrics Benchmarking Report: Preliminary Findings and Recommendations (October 2010)," prepared by Plexus for the Internal Revenue Service (IRS), IBM's Global Business Solutions and Learning and Knowledge Organizations are trying to focus on the knowledge sharing model through social solutions, as shown in Figure 1.1, to derive their business values.

The Barr Foundation in Boston has quarterly reflection meetings where program officers are asked to reflect on extreme successes, failures, or surprises during the past quarter in terms of what they learned as related to their program goals and objectives. "Extreme" events are used as it is felt that the most learning would take place from these situations. The Barr Foundation then has half-day to full-day reflection meetings with the staff to share and capture these "aha" moments.

The Skoll Foundation, Chevron, the World Bank, and other organizations also include learning and knowledge sharing proficiencies as part of their annual employee performance review. In this way, people are more apt to collaborate and share knowledge if they are recognized and rewarded for these types of behaviors.

At NASA Jet Propulsion Laboratory (JPL), and throughout the NASA centers and NASA headquarters, storytelling is actively used as a way to share knowledge. According to JPL (http://km.nasa.gov/home/km_jpl.html),

Professional Development	Productivity
• Increase visibility, recognition and reputation in organization • Foster personal connections and grow their personal networks • Promote continuous learning/knowledge sharing culture	• Accelerate time to locate & access expertise • More rapid identification of people who can positively influence business outcome • Increase opportunities for innovation • More expedient knowledge creation & sharing • Reduce time to perform activities
Knowledge Sharing	**Collaboration**
• Increase awareness and leverage of expertise in the business as it evolves • Increase x-department/x-geo collaboration • Accelerate pervasive dissemination of knowledge (codified and tacit) • Optimize the use of content through social networks	• Increase amount of informal and formal cross department & cross geo collaboration • Visibility of formal and informal communities –information flow/collaboration & health of network • Visibility of expertise & faster reciprocal contact due to social network introductions • Increase efficiency and effectiveness of collaboration

FIGURE 1.1
IBM's Global Business Solutions/Learning and Knowledge Organization proposed value matrix for knowledge sharing.

The JPL Library hosts a monthly forum where a JPL person tells a story surrounding a mission event. The focus is on moving from personal knowledge to organizational knowledge, individual to collective knowledge. The forums emphasize sharing one's identity, building relationships, bonding, community, and creating a knowledge-sharing culture through socialization of new members, mentoring, and lessons learned.

Probably the most common application of knowledge sharing in organizations is through the use of online communities of practice. Most organizations use online communities to share knowledge, reach out to others, post ideas and documents, ask questions to the community, capture knowledge, and encourage collaboration to innovate. Shell Oil, for example, uses LiveLink and SharePoint for online communities, along with other collaboration tools and Shell Wikis. Fluor Corporation, Hershey's Corporation, Best Buy, Cisco, PricewaterhouseCoopers, The Aerospace Corporation, World Bank, Pfizer, Food and Drug Administration, Michelin North America, and thousands of other organizations worldwide are using online communities to capture, share, apply, and generate knowledge.

HOW CAN WE IMPROVE THE STATE-OF-THE-ART IN KNOWLEDGE SHARING

According to Wallace et al.'s 2011 article, "The Research Core of the Knowledge Management Literature" (*Int. Journal of Information Management*, Vol. 31), they found that about two-thirds of the knowledge management research methodologies are either nonexistent or ad hoc. From their bibliometric and content analyses on knowledge management literature of 21,596 references from 2,771 source publications, 27.8% used no identifiable research methods. Of the remaining refereed articles, 60% employed mainstream social sciences research and 40% used provisional methods as a substitute for more formally defined or scientifically based research methodologies. From these statistics, the knowledge management community could certainly strengthen the rigor of the methodologies and techniques being applied in the KM field.

Another growing area of concern deals with cross-generational knowledge flows. As the baby boomers continue to retire and more Gen Y'ers enter the workforce, more focused attention needs to be made on how best to capture, share, transfer, and leverage the knowledge across the organization's generational workforce. From my research in this area, we found that there are five critical success factors for cross-generational knowledge flows (although, these can apply even within generations): shared values, reciprocity (in terms of equal sharing of knowledge when needed), intrinsic worth of the knowledge, convenient knowledge transfer mechanism, and interpersonal trust and respect. Organizations will continue to deal with issues relating to onboarding, knowledge retention, decision making inclusive of all generations, and the like. Also, the younger workforce are digital natives and take collaboration and social networking techniques for granted. Thus, management will have to find ways to best tap and utilize these skills for knowledge creation activities and building a strong sense of community within the organization.

As global competition and recessionary periods remain prevalent, organizations will have to apply knowledge management, business intelligence (BI), and competitive intelligence (CI) techniques to improve the "strategic intelligence" of the firm. In other words, the techniques used in KM, BI, and CI can be leveraged to enhance the organization's strategic decision-making ability. Further development and application of these techniques for optimizing both the internal and external intelligence of the firm will need to continue.

As mobile computing, personalization, cloud computing, and Web 2.0 continue their bold pace, organizations will have to see how knowledge management can best take advantage of these trends. Certainly, these advances will affect collaboration and social networking, and organizations will need to rethink how best to utilize these developments. Related to these issues, organizations will continue to be concerned with cybersecurity and caring for their organizational assets. Thus, privacy, computer and individual security, ethics, and legal issues will continue to be on the minds of organizational leaders.

A last area on which the knowledge management community should focus its attention deals with metrics and ways to measure the benefits of knowledge sharing initiatives. Too often, organizations are using mainly systems and output measures for knowledge management. Rather, outcome measures should really be at the heart of the organization in terms of, for example, how we can measure the impact of knowledge sharing initiatives in terms of innovation, profits, and customer perceptions. Metrics to better measure the alignment of knowledge management activities with the strategic goals of the organization will continue to be necessary in the future, especially as we will need to do more with less based upon resource constraints. Certainly, techniques like value network analysis, activity-based costing, and the like could shed additional light on these measures.

WHAT IS AHEAD IN FORTHCOMING CHAPTERS

The *Knowledge Management Handbook* was the first of its kind and was published in 1999 by CRC Press. Since then, there have been a number of excellent handbooks in knowledge management by others. Through the urging of CRC Press, we thought we would publish a second edition of the *Knowledge Management Handbook* under the condition that we would focus on two of the key elements of knowledge sharing—that is, collaboration and social networking.

With that goal in mind, this handbook represents cases, applications, concepts, techniques, methodologies, issues, and trends as related to collaboration and social networking in a knowledge management context. Many of the chapters represent leading authors and organizations worldwide, and the hope is that this handbook will serve as a key reference source, as well as a classroom text, for those engaged in knowledge management, especially from a collaboration and social networking perspective.

The times ahead look promising for knowledge management, but we must continue to apply the science behind the art. As the field continues to mature and KM is woven within the organizational fabric, gains from KM initiatives will continue to heighten. Already, for example, Linda Davies, the Director of Knowledge Communications at Mars, Inc., reported at the United Kingdom KM Conference in June 2011 that their KM initiatives at Mars, mainly through networks and communities of practice, are expected by the end of 2011 to have a measured $1 billion bottom-line value to the business (Mars started its official KM journey in 2004). The times ahead for knowledge sharing look exciting indeed!

2

Knowledge and Collaboration in Multihub Networks: Orchestration Processes among Clinical Commissioning Groups (CCGs) in the United Kingdom

Celine Miani, Markos Zachariadis, Eivor Oborn, and Michael Barrett

INTRODUCTION

Innovation networks are often characterized by loose, semitemporal link-ages between actors who seek to employ the right resources and engage in strategic collaborations in order to deal with specific problems and develop innovative services and solutions (Van Wijk et al., 2003). Due to the absence of tight structures and hierarchical authority, these networks usually rely on autonomous hub entities or strategic centers to maximize their efficiency and support their innovation output objectives (Hagel et al., 2002). In this context, the role of the hub firm is to coordinate, direct, and influence network members through a number of "orchestration pro-cesses" that will enhance innovation development and will ultimately create and extract value from the network (Kogut, 2000; Dhanaraj and Parkhe, 2006). These processes, which have employment and effectiveness that may vary according to the context and the characteristics of the inno-vation network, involve the management of knowledge mobility, inno-vation coherence, network membership, network stability, innovation appropriability, and innovation leverage, among others (Van Wijk et al., 2003; Nambisan and Sawhney, 2011). In this chapter, we concentrate on four of the above that are more relevant to our case study on the health-care sector: managing knowledge flows, managing innovation coherence, managing network membership, and managing network stability.

Until recently, these management practices remained largely underin-vestigated, having implications for our understanding on how innovation networks mobilize knowledge and facilitate collaborations between members. In addition, most of the existing studies and respective examples (e.g., Iansiti and Levien, 2004; Nambisan and Sawhney, 2007, 2011) focus on the technology and manufacturing sectors (e.g., the cases of Boeing, Apple, and Facebook), leaving service innovation networks in other parts of the economy relatively unexplored. To fill that gap, our study draws from the healthcare sector in the United Kingdom and examines how medical innovation and healthcare commissioning practices progress under the direction of multiple clinical commissioning hubs (CCGs) that act as network orchestrators.

Over the last couple of decades there has been tremendous growth on the study of social networks with scholars converging from different fields such as economics, sociology, anthropology, organizational behavior, and other social sciences to study the "pattern of interconnections among a set of things" (Easley and Kleiberg, 2010, p. 1). Much of this research examines the transfer of knowledge through ties (Van Wijk et al., 2003). Given that knowledge is seen as an essential component of innovation and a strategic resource (Grant, 1996), increased knowledge sharing is often associated with enhanced organizational performance (Aral et al., 2006). For that reason its mobilization should be a priority in every orchestration plan. Nevertheless, social network analysis (SNA) has shown that knowledge flows are also subject to a number of network characteristics like network structure (Granovetter, 2005; Burt, 2004), the actor's network position (Uzzi, 1996), and the means of communication between network ties (Wu et al., 2008). All of the above can influence the ease with which knowledge is being diffused and utilized within the network. In that context, hub firms can effectively manage knowledge flows and ensure access to information that resides in different locations in the network by taking advantage, for example, of the centrality and status of actors, the autonomy of the net-work, and the technological infrastructures in place. As a consequence, actors do not need to sacrifice any resources to create knowledge and build capabilities that can be found elsewhere in the network (Hargadon and Sutton, 1997). Network orchestrators can achieve this through a number of activities including the deployment of technologies to identify, absorb, and utilize knowledge, as well as the arrangement of events to encourage socialization between members (Dhanaraj and Parkhe, 2006).

CLINICAL COMMISSIONING HUBS (CCGS) AS INNOVATION NETWORK ORCHESTRATORS

Figure 2.1 illustrates the framework used in our case study to examine how CCGs draw from the characteristics of their network (on the left) in order to facilitate the various orchestration processes (on the right) and maximize their innovation output.

Knowledge sharing also plays a crucial role in ensuring innovation coherence—by that we mean the alignment and relevance of the innovation goals and visions with the broader market environment (externally), and within the network itself (internally) (Nambisan and Sawhney, 2011). This is particularly important when it comes to healthcare innovation where clinical pathways need to not only reflect the latest technologies but also reproduce developments in medical research. Within this context, innovation network orchestrators play a strategic role in managing external and internal innovation coherence by identifying trends in research and

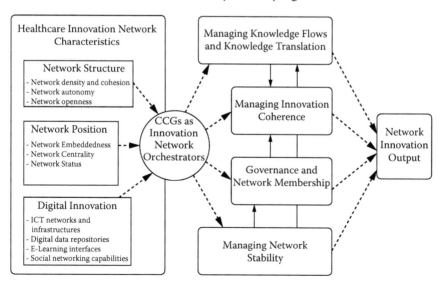

FIGURE 2.1

Innovation network orchestration in action: clinical commissioning hubs (CCGs), orchestration processes, and network characteristics.

technology and coordinating activities within the network. Network openness as well as high network density* and cohesion can assist in that direction. It is expected that lack of internal innovation coherence can lead to increased costs through unnecessary specifications, incompatibilities, delays, and even duplications in the process (Gerwin, 2004). In addition, the absence of external innovation coherence will directly undermine the value of the innovation output, as it may not be relevant with current developments.

As the size of the network becomes larger, the responsibility to coordinate activities becomes more demanding and crucial. In that respect, network membership should be managed carefully because it not only influences the innovation output through the diversity of the actors but also has a direct effect on the innovation coherence and the stability of the network. Even though the diversity and size of the network can be considered as taken-for-granted network design characteristics (Dhanaraj and Parkhe, 2006), hub organizations can often influence the degree of membership through effective governance of their alliance network. Recruiting participants should be done considering the benefits from access to capabilities and brokerage opportunities with other interactions (Gimeno, 2004) that will increase innovation output. Subsequently, the change in strategic partners and the introduction of new players can affect the structure of the network (e.g., its density and openness), as well as the network positions of the existing actors (e.g., their centrality and status). Hence, managing network membership becomes even more vital as it can significantly alter the network architecture.

Finally, network stability is also crucial. Unstable networks can impact knowledge flow and risk the promotion of new collaborations or alliances. In the face of uncertainty, actors may move to competing networks or withhold valuable information by creating cliques that serve the narrow interests of local members. Dhanaraj and Parkhe (2006) highlight three ways through which hub firms as orchestrators can manage the stability of their network: enhancing reputation, anticipating future benefits, and building multiple joint projects. The three are interlinked as the higher the reputation of market leadership of the hub, the more joint projects will be created reinforcing forward-looking expectations for members and

---

* We need to keep in mind that innovation networks are usually low-density/high-centrality networks. In these networks high centrality does not necessarily imply power and authoritative control, as these are usually "loosely coupled" (Dhanaraj and Parkhe, 2006, p. 660) systems of independent and autonomous firms.

therefore making the network more resistant to dissolution (Kenis and Knoke, 2002). Managing the stability of the network becomes even more critical in periods of transition and change like the one we examine in our case study.

As can be seen, all the above processes have a direct effect on the innovation output of the network and thus, the value that this produces. Different processes can affect one another directly but also indirectly via the influence some of them can have on the network structure or the actors' positions. In that respect it is the responsibility of the hub firms—in our case the CCGs—to manage these collectively and take into account the interactions between them and the network characteristics.

CASE STUDY BACKGROUND: DEVELOPMENT OF (CCGS) COMMISSIONING AT NATIONAL HEALTH SERVICE (NHS)

Following the announcement of the latest UK National Health Service (NHS) reform (Department of Health, 2010), the health system in the United Kingdom has entered a new cycle of radical changes that aim to improve healthcare outcomes and increase efficiency. At the center of the strategy proposed by the current government is to "liberate the NHS" and put clinicians "in the driving seat and set hospitals free to innovate, with stronger incentives to adopt best practice."* Along these lines, the reform challenges the way commissioning of healthcare services is organized and executed.

Over the years, the sustained pace of medical innovation combined with an aging population with increased demand has put more pressure onto the health system. Modern health economies are thus seeking new ways to manage knowledge and innovation, and share best practices quickly and efficiently. In that context, networks turn out to be of particular relevance and increasing importance as they allow organizations and their employees to access new ideas and innovate. The importance of networks in developing new clinical practices and diffusing healthcare innovation was discussed in previous studies that elaborate on the role of collaboration

---

* See the white paper and the subsequent Health and Social Care Bill produced by the Conservative-Liberal Democrat administration as part of their health policy in England and Wales (Department of Health, 2010).

and organizational learning in efficiency gains (Fleuren et al., 2004; D'Amour et al., 2008). In line with this research, the latest reorganization of the NHS in the United Kingdom provides us with a unique opportunity to explore how knowledge is being shared and managed among healthcare networks in order for general practitioners (GPs) to develop their capacity to make commissioning decisions.

The recent restructuring introduced a new type of assembly called clinical commissioning groups (CCGs), which are essentially clusters of physician-led primary care practices responsible for the procurement of healthcare services for a given population. Commissioning decisions are often time-consuming and demand particular knowledge and competencies that are not part of the conventional attributes of GPs; nevertheless, under the new scheme they are assumed to take control and explore new possibilities in order to develop novel commissioning activities. This exercise demands a transition period during which new commissioners need to build their expertise on commissioning issues, manage and share their knowledge, collaborate with colleagues and external stakeholders, and seek advice from peers in different clusters or within their groups. Throughout this stage, we were able to observe some groups and monitor the network development of GPs and managers involved in this process.

Influences and Previous Initiatives

Despite the fact that the new bill has been described by many as disruptive and revolutionary, GP commissioning is not entirely new. A similar scheme was tried and tested successfully during the 1990s when, according to claims, "GP fund holders" were able to improve healthcare delivery locally.* However, the scheme was abandoned when the Labor party came into power again in 1997. The stated reasons for this about turn were that not all patients received equal access to healthcare—as not all GPs took on Fundholding status—and overall costs were rising. In spite of the ongoing criticism, the subsequent (Labor) government appreciated the merits of GP-led purchasing and eventually formed a new scheme where "local

---

* GP budget holders were introduced originally in 1991 as part of the "Working for Patients" white paper of the Conservative government (Department of Health, 1989). An extension of the GP fund-holding scheme called "Total Purchasing" was also piloted in 1995 to allow general practitioners to buy all the hospital and community health services for their patients in collaboration with the local authorities.

GP-led commissioning groups" would take over from the existing GP budget holders and health authorities—the local NHS administrative unit. The new systems' aim was to reduce inequality and lower management and transaction costs, but as the new primary care groups (PCGs) grew bigger and were eventually transformed into primary care trusts (PCTs), GPs gradually lost influence in the management of the organizations. To deal with the increasing bureaucracy and inefficiency of PCTs, the government introduced a new form of practice-based commissioning (PBC) that would reengage GPs whose involvement would create "a greater alignment between financial consequences and the decisions to refer, and [would bring] the market closer to patients" (Brereton and Gubb, 2010, p. 65). Eventually, PBC paved the way to the groups that this study investigates (Figure 2.2). An ongoing theme in the evolution of the GP commissioning process over two decades is the changing policy context and creeping encroachment of bureaucratic inefficiency.

Overall, by looking at the content and outcomes of those past commissioning experiences, one can easily discern the benefits that such initiatives brought into the context of healthcare. Studies show that such schemes had positive effects globally by shortening waiting time for patients, reducing rates of hospital admissions, and containing the rise of prescribing costs (Lewis, 2004). Unfortunately, several of these studies lack rigor because of the difficulty of comparing different healthcare policies and schemes. In most of the cases, proposed systems were not implemented for a sufficiently long duration, as well as lacked adequate and accurate data, to draw robust conclusions.

In addition to the local UK strategies and expertise, ideas and inspiration for the British policies were often taken from the other side of the Atlantic. Since the 1980s, a form of GP-led commissioning developed in the United States, and more specifically in California. There are the independent practice associations (IPAs) and large medical groups arrangements with health plans. Funding, payment, and management modes

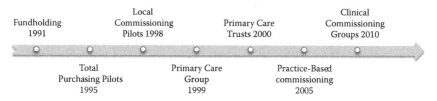

FIGURE 2.2
Clinical commissioning timeline 1991 to 2011.

can vary, but both remain roughly GP organizations run by GPs. Among those groups some have been closely studied by Casalino (1997) who demonstrated that such structures are "a new kind of organization, with new capabilities," and "an opportunity for innovation and improvement."

Current Reform and Challenges

Building on the implementation processes of previous policies, the last reform pursued by the coalition government essentially proposed the following:

- Clinical commissioning groups (CCGs) are GP-led groups with the responsibility to design, negotiate, and purchase most of the healthcare services for the population they cover. All GP practices must belong to a CCG.
- CCGs need to be authorized by the new NHS commissioning board.
- Primary care trusts (PCTs) and strategic health authorities (SHAs) are to disappear. (They are, respectively, the local and regional administrative levels of the NHS.) These are being replaced by conjoining and federating the SHAs and PCTs into new, more centralized administrative bodies.
- Health Watch units will serve patient interests at the local level, while local health and well-being boards will bring together councillors, key commissioners, public health directors, social services, and Health Watch units.

Each CCG will have its own structure and organization, inherited from previous policies and local dynamics. But in any case, CCGs share some common ground when it comes to their general shape: at the cluster/local level, which may correspond to a former PBC group, there are a number of practices whose leads are collaborating in order to resolve local issues. The cluster leads are also part of the CCG board, which makes commissioning decisions for a wider area and contract with providers. All those meetings and pursuits are conducted in close relationship with the PCT.

It is evident that despite the potential benefits, GP-led commissioning schemes face limitations and overwhelming challenges. In line with other studies, the Nuffield Trust (Thorlby et al., 2011) reported that the main weakness of medical groups is the high probability of financial failure. Moreover, such organizations demand strong leadership capacity in order to negotiate tariffs of services and contracts and noteworthy investments in information technology (IT). A study evaluating the total purchasing Pilots (Goodwin et al., 1998) shows that the organizational investment to be made in order to reach objectives is significant: only groups that heavily invested in management and organizational services were successful. A study by the King's Fund (Curry et al., 2008) regarding the PBC reveals similar concerns about the lack of data quality, organizational instability, weak engagement of frontline GPs, and the unwillingness of PCTs to "let go." All those challenges contribute to the significant criticism the current reform has received since it was announced in 2010.

NETWORK MANAGEMENT STRATEGIES AND PRACTICES

Even though discussing the validity of the reorganization is not our goal, we explore the influence of knowledge and networks on the innovation outcome. In a transition, hierarchies change and collaborations are threatened or reinforced; thus the dynamics of the relationships can alter significantly. This fact, combined with the urgent requirement of creating new and efficient organizations, means that CCG groups need to develop orchestration skills to manage knowledge flows, innovation coherence, network membership, and network stability.

Managing Innovation Coherence

In the changing context of the reform, managing innovation coherence becomes critical: the need to organize networks and work collaborations in a relevant and cohesive way to motivate and spread innovation requires a certain hindsight and strong orchestration.

External Innovation Coherence

The external coherence (i.e., the coherence of the CCG activity compared to the national trend) is influenced by top-down political pressure and numerous clinical guidelines coming from regional and national NHS bodies. At a more local level, coherence is orchestrated by the PCTs, which share common practices and skills with peer PCTs across the country. PCTs also fund CCG training and hire the services of consulting organizations that go to CCGs and organize local workshops, participating in the dissemination of innovation that supports the creation of a common knowledge ground across CCGs. Overall, this creates a shared learning throughout England without the different groups being actually connected to each other. Initiatives come also from the CCGs, as they seek to find out what other CCGs do and gather knowledge from various sources. However this last strategy remains quite limited as the CCGs are currently internally focused. In addition, leading members of CCGs lack time as most are also engaged in regular patient contact, the primary source of income for their practices. One of them reports that "if there is anything in the press obviously I will raise it, but I don't have time."

Internal Innovation Coherence

The more significant orchestration processes in terms of innovation coherence may occur internally. CCGs leaders are trying hard to build a vision and global strategy for their group. Having such a clear view of the objectives and development goals, "something a bit more meaningful that is going to get all these (GP) practices to say yes I want to join you as a consortium" (as suggested by a CCG manager) should provide the organization with internal innovation coherence. This vision translates into the development of terms of reference which are basic membership requirements as to which practices sign up. This legal binding, plus the peer pressure (enhanced by monitoring at the practice level) increases the coherence of the changes. Standardized IT practices and resources constitute the most practical aspect of innovation coherence; they concretely create coherence by harmonizing data and making medical records more accessible. Finally, strong leadership reinforced by communication of the best practices for knowledge sharing between the different levels of the CCGs ensures that the innovation coherence is not only supported but also spread harmoniously across the organization. As a GP noted, a pressing goal is to "make sure that the way that the information is disseminated back and forth works."

Managing Network Membership

In addition to CCG network membership being facilitated through legally bound contracts, some groups have membership fees and others have organized yearly forum meetings. These practices create a strong sense of belonging to an organization, through networking, information flows, and sharing of ideas. On a more regular basis, network membership is also facilitated by the dissemination of newsletters, leaders' visits, and local meetings. CCG members in particular acknowledged the value of local GP practice visits:

> We do (GP) practice visits […], we tend to go around and see them and we find that the most productive way of actually communicating with practices is actually to go out to them and have an agenda and sit down with them.

Further, the engagement of the front-line GPs facilitated the implementation of clinical guidelines, and the fact that they share resources—IT systems or administrative staff—further facilitates the link between different practices or clusters across the CCG.

Finally, the mandate of leaders and the fact that they are elected by their peers give them both legitimacy and a strong sense of belonging to the CCG. A GP commented:

> I think the negotiators consider themselves to be representing the group, and the group consider themselves to be representing the GP practices, the clusters and the GP practices. I do believe that to be true, the importance of the elections is so that it can be shown to be true, if that makes sense, that they actually have.

The fact that they are accountable at different levels, to the practices as GPs, to the clusters as lead, to the CCG as executive members, and to the regional NHS as commissioners, strengthens the network membership by adding layers of affiliation.

Managing Network Stability

Governance

Considering the costs of reorganization and the potential loss (both financial and human resources [HR] related) due to instability, managing network stability now and securing it in the long term is a key factor of CCG success. A PCT manager pointed out:

> You have to agree the constituencies and agree all of the pre-election machinery, so how many representatives there are going to be, what the constituents are going to be, who's entitled to vote, how long are they going to serve for, what's the commitment, role specifications.

Such stability cannot be reached without putting a significant amount of effort and resources into assessing the strength of the emerging organization through solid governance of alliances and collaborations. This requires both the design of accountability mechanisms and the development of the CCG function, trust, and effective leadership through the establishment of good reputation and reassurance of the network in the future.

Relationships

Our case suggests that the appropriate distribution and clarity of the roles that employ the right resources and capabilities in the network in the best way create maximum value (Kogut, 2000). Further, network stability was seen to result from building synergistic relationships with PCTs so as to retain the workforce with critical commissioning skills, thus enabling a form of organizational memory to be developed within the network. One useful example of implementing national guidelines efficiently was seen in a CCG that established close collaboration between a clinician and a PCT manager. This pair teamed up on a specific work stream (e.g., procuring effective services for old-age care and emergency care) and also developed specific in-house competencies (i.e., contracting), thus maximizing the impact of the CCG by using all relevant resources. Importantly they sought to seamlessly combine clinical input with financial knowledge without wasting administrative and human resources, thereby contributing efficiently to the network stability. In this case there was no need to outsource, but rather they worked hard to develop cooperative mechanisms within the current network, thus mobilizing the available knowledge and "combining relevant technologies in novel ways" (Dhanaraj and Parkhe, 2006, p. 662).

Finally, efforts must be developed in building trust and long-lasting partnerships with a diversity of external stakeholders (external stability) and nurturing trusting relationships within the CCG by securing engagement of the leads and engagement of the front-line GPS (internal stability). A GP described the process:

> I suppose if it's new people you are working with it's about building up that trust and, you don't have to know everything they are doing because you then know they are doing it in the way that you are happy with, so I think that takes a bit of time.

Managing Knowledge Flows

Effective knowledge sharing is critical in facilitating the practices and strategies described above.

Knowledge Brokering

Effective knowledge sharing between members must be ensured in order to maximize the innovation output. Organizational boundaries within and between organizations can jeopardize access to knowledge that resides in different locations in the network. It should be the responsibility of the network orchestrator to ensure knowledge mobility that will connect existing ideas with potential problems and will create efficiencies within the system. Hargadon and Sutton (1997) call this activity "technology brokering." Yet, the institutionalization of knowledge sharing through the close cooperative knowledge sharing between clinical and administrative pairs will remain limited, as in our findings, if no one takes ownership of coordinating (but not controlling) the whole process. As in every network, the CCG hub requires key players who will provide "subtle leadership" (Orton and Weick, 1990), encourage knowledge mobility, and enhance innovation activities. In the absence of strict hierarchies, those leaders need to develop brokering strategies and facilitate links and collaborations between stakeholders. A PCT director describes his brokering role in the new commissioning framework:

> I see my role as helping them to come up with plans, commissioning plans and then unlocking the resources here to help to deliver it. ...One of the things I can do and the other directors do in their equivalent roles, is broker and unlock, or try to anyway, the resources that they need, the people usually that they need to implement the plans that they have come up with.

A mechanism of in-house locally directed feedback and upstream external transmission of ideas is also necessary to keep network members up to date and ensure relevant follow-up of commissioning and policy developments. Considering the importance of the brokers (who may be GPs, practice managers, or PCT directors), it was judged relevant to develop personal coaching and training sessions to improve their individual performance as well as that of the group. During the initial stages of the process that we examined, PCTs tended to organize training sessions for the clinicians to facilitate an understanding of the functions of commissioning. Yet in most cases GPs acknowledged that there was little value from these sessions. The PCTs have subsequently come to realize that clinicians do not need to know the majority of the daily job details but must rather strategically manage knowledge mobility. The emphasis is not so much on transferring content but managing knowledge flows and brokering. A PCT director put it this way:

> It's not so much about training so much as emotional intelligence, management, experience and skills. It's not just concrete knowledge, it's softer skills around you influencing, managing.

Digital Networks

In the new health context, knowledge flows involve more than clinical content requiring the sharing of ideas and practices in larger and more generalist networks. As a result, the current SHA is building up an interactive learning network for CCGs which has been designed by an external software vendor. The Web site is made up of videos, interviews, and case studies that aim at disseminating good practices and useful information. Such electronic networks often provide incentives that meet the reputation argument for motivating participation (Wasko and Faraj, 2005). However, despite the increasing influence of social networking platforms, GPs have in general made limited use of them to date. They remarked that they do not have time to engage with the material, or that they prefer to hear the information from people they trust and with whom they share a relationship. An active member of the British Medical Association (BMA) suggested that they would rather read the BMA generalist newsletter than spend time on a learning network, even though the learning network may be more relevant to his commissioning needs. People feel the need to share locally and to be close to knowledge sources. A frequent finding was that CCG members expected the leader to give them a summary of the information found on the Internet and other electronic network sites. A GP practice manager mentioned:

> It's easy to say circulate round the doctors, but they won't read it. That's not the same as informing them. I usually grade [the e-mails] and in the subject heading add—read this one, [...] so give them an idea when they're skimming through.

Despite the current limited adoption of IT use by GPs and the seeming lack of wider acceptance, GPs however pointedly noted the need for more IT capability in producing analytics for the commissioning process. This stems from a perceived lack of basic costing, quality outcomes, and general service provision details being available to inform the commissioning process. This underscores the importance of digital networks and infrastructure in complementing and fostering network orchestration. Some mentioned that what they need is a "Google of commissioning": a research tool where they could search for key terms as "commissioning mental health" and receive all the related information. Along these lines, most of the groups now have dedicated Web sites with public information sharing all the commissioning activities of the groups. Apart from knowledge sharing this also underlines their accountability to the public and encourages patient participation to provide feedback on their services, an important aspect of their patient service provision and user-generated knowledge.

Furthermore, in order to be effective in commissioning, GPs need to have information and knowledge on prices, outcomes, tariffs, and returns. In that context, data and IT systems can be seen as facilitating knowledge brokering across boundaries, including between the hospital and the practice, between social and health care, and between the specialist and the GP. As such, medical records need to be more widely shared electronically. But it can go across and beyond that. The clinical and financial data accessible from a centralized repository can enable improved clinical management in patient related decisions. A GP commented:

> Clinical coding has to be accurate and commensurate with what has happened to the patient [...] and there has been a number of situations where that doesn't appear to have been the case and things are charged at a higher rate than they should be or more procedures are charged for them than actually took place [...] we have got to get it as accurate as we can so that we can use it both to audit and analyse and also to plan ahead.

Clinical groups are asking for a nationally uniform and integrated database, while starting to develop equivalent initiatives at the local level. In reality, even at the local level, GPs and the local providers find they are at odds on a variety of data issues such as inconsistent referral figures. The inability to access accurate and consistent data is a significant threat to managing knowledge flows in the commissioning process.

CASE IMPLICATIONS FOR THE FUTURE

Hub groups have the responsibility to integrate and manage innovation production activities through a number of processes. However, our case study suggests that most actors do not realize the potential benefits they could gain by taking advantage of the networks they belong to: some networks are tacit, some are unused. Only a few hubs have optimized the relationships they created to reach their new goals. In the current transition toward CCG and a new commissioning structure, effective leadership and an appropriate structure are needed in a network-centric innovation setting. The network and innovation orchestrators are critical in the success or failure of a CCG in taking over commissioning duties. The effectiveness of the network and the innovation output will depend on their commitment as well as the intensity of their ties with colleagues, partners, and stakeholders. These ties in turn are needed to facilitate the mobility of knowledge for collaborative leadership involving a range of possible incentive schemes for engagement in the reform and to reward proactive orchestrators financially, politically, and through in-kind mechanisms.

REFERENCES

Aral, Sinan, Erik Brynjolfsson, and Marshall Van Alstyne. 2006. Information, technology and information worker productivity: Task level evidence. In W. Haseman, D.W. Straub, S. Klein, and C. Washburn (eds.), *Proceedings of the 27th International Conference on Information Systems*, Paper 21. Milwaukee, Wisconsin.

Brereton, Laura, and James Gubb. 2010. *Refusing Treatment: The NHS and Market-Based Reform*. London: Civitas.

Burt, Ronald S. 2004. Structural holes and good ideas. *American Journal of Sociology* 110: 349–399.

Casalino, Lawrence, and James Robinson. 1997. The Evolution of Medical Groups and Capitation in California. Report for the Kaiser Family Foundation, Washington, DC.

Curry, Natasha, Nick Goodwin, Chris Nayllor, and Ruth Robertson. 2008. *Practice-Based Commissioning: Reinvigorate, Replace or Abandon?* London: The King's Fund.

D'Amour, Danielle, Lise Goulet, Jean-François Labadie, Leticia San Martín-Rodriguez, and Raynald Pineault. 2008. A model and typology of collaboration between professionals in healthcare organizations. *BMC Health Services Research* 8(January): 188.

Department of Health. 1989. *Working for Patients* (Cm. 555). London: HMSO.

Department of Health. 2010. *Equity and Excellence: Liberating the NHS*. London: HMSO (http://www.dh.gov.uk/en/Publicationsandstatistics/Publications/Publications PolicyAndGuidance/DH_117353).

Dhanaraj, Charles, and Arvind Parkhe. 2006. Orchestrating innovation networks. *Academy of Management Review* 31(3): 659–669.

Easley, David, and Jon Kleinberg. 2010. *Networks, Crowds, and Markets: Reasoning about a Highly Connected World*. Cambridge, UK: Cambridge University Press.

Fleuren, Margot, Karin Wiefferink, and Theo Paulussen. 2004. Determinants of innovation within health care organizations: Literature review and Delphi study. *International Journal for Quality in Health Care* 16(2): 107–123.

Gerwin, Donald. 2004. Coordinating new product development in strategic alliances. *Academy of Management Review* 29(2): 241–257.

Gimeno, Javier. 2004. Competition within and between networks: The contingent effect of competitive embeddedness on alliance formation. *Academy of Management Journal* 47(6): 820–842.

Goodwin, Nicholas, Nicholas Mays, Hugh McLeod, Gill Malbon, and James Raftery (on behalf of the Total Purchasing Evaluation Team, King's Fund). 1998. Evaluation of total purchasing pilots in England and Scotland and implications for primary care groups in England: Personal interviews and analysis of routine data, *BMJ* 317 (7153): 256–259.

Granovetter, Mark. 2005. The impact of social structure on economic outcomes. *Journal of Economic Perspectives* 19(1): 33–50.

Grant, Robert M. 1996. Prospering in dynamically-competitive environments: Organizational capability as knowledge integration. *Organization Science* 7(4): 375–387.

Hagel, John III, Scott Durchslag, and John Seely Brown. 2002. Orchestrating Loosely Coupled Business Processes: The Secret to Successful Collaboration. Working paper.

Hargadon, Andrew, and Robert I. Sutton. 1997. Technology brokering and innovation in a product development firm. *Administrative Science Quarterly* 42: 716–749.

Iansiti, Marco, and Roy Levien. 2004. *The Keystone Advantage*. Boston: Harvard Business School Press.

Kenis, Patrick, and David Knoke. 2002. How organizational field networks shape inter-organizational tie-formation rates. *The Academy of Management Review* 27(2): 275–293.

Kogut, Bruce. 2000. The network as knowledge: Generative rules and the emergence of structure. *Strategic Management Journal* 21: 405–425.

Lewis, Richard. 2004. *Practice-Led Commissioning: Harnessing the Power of the Primary Care Frontline*. London: The King's Fund.

Nambisan, Satish, and Mohanbir Sawhney. 2011. Orchestration processes in network-centric innovation: Evidence from the field. *Academy of Management Perspectives* (25)3: 40–57.

Orton, J. Douglas, and Karl E. Weick. 1990. Loosely coupled systems: A reconceptualization. *Academy of Management Review* 15: 203–223.

Thorlby, Ruth, Rebecca Rosen, and Judith Smith. 2011. *GP Commissioning: Insights from Medical Groups in the United States*. London: Nuffield Trust.

Uzzi, Brian. 1996. The sources and consequences of embeddedness for the economic performance of organizations: The network effect. *American Journal of Sociology* 61(August): 674–699.

Van Wijk, Raymond, Frans A.J. Van Den Bosch, and Henk Volberda. 2003. Knowledge and networks. In *The Blackwell Handbook of Organizational Learning and Knowledge Management*, ed. Mark Easterby-Smith and Marjorie A. Lyles, 428–453. New York: Wiley-Blackwell.

Wasko, Molly McLure, and Faraj, Samer. 2005. Why should I share? Examining social capital and knowledge contribution in electronic networks of practice. *MIS Quarterly* 29(1): 35–57.

Wu, Lynn, Ben Waber, Sinan Aral, Erik Brynjolfsson, and Alex Sandy Pentland. 2008. Mining face-to-face interaction networks using sociometric badges: Predicting productivity in an IT configuration task. R. Boland, M. Limayem, and B. Pentland (eds.). In *Proceedings of the International Conference of Information Systems*, Paper 127. Paris, France.

3

Religious Communities of Practice and Knowledge Management—The Potential for Cross-Domain Learning

Denise A.D. Bedford

INTRODUCTION

According to Jean Lave and Etienne Wenger,[1] a *community of practice* is a group of people who share an interest, a draft, or a profession. It is a group that evolves naturally because of the members' common interest in a particular domain, or it can be created specifically with the goal of gaining knowledge related to their field. It is through the process of sharing knowledge and experiences with the group that the members learn from each other, construct a common practice, and build tight bonds of community.[2,3]

Communities of practice play a well-recognized and well-documented role in the creation, exchange, mobilization, and capture of knowledge. John Seely Brown[4] was among the first to document the existence of informal communities of practice and to describe the value of these communities to organizational knowledge management. Wenger, McDermott, and Snyder[5] discussed community components and life cycles and suggested ways that community members can facilitate the development of communities of practice. Saint Onge[6] further discussed the role that organizations can play in supporting structured or more formal organizational communities of practice. The literature is rich with case studies of communities of practice in a wide range of subject domains and economic sectors, including military services,[7-9] national intelligence, diplomatic service, disaster response,[10-13] health care,[14-23] wilderness medicine,[24] technology support, software development, education,[25-29] and many others. Some research has begun to focus on the role of communities of practice beyond formal organizations and formal business roles, for example in craft and hobby environments,[30-32] community writing centers,[33] and online gaming communities.[34]

BODY OF KNOWLEDGE ON COMMUNITIES OF PRACTICE

As a result of the longstanding focus on communities of practice, there is a rich body of knowledge about factors that lead to their success or failure. We know that healthy communities of practice are designed for growth and evolution—they are dynamic not static. Healthy communities have built-in opportunities for dialogue within the community and between the community and its larger environment to ensure there are continuous opportunities for growth and review. Within a sustainable community there are different levels of participation—from active participation, to occasional participation, to observation status. Most communities of practice, though, have an active and passionate core group of members. Healthy communities also have both public and private spaces for members to develop their practice. Perhaps most important, sustainable communities focus on the value the practice gives to the community members, and they build upon and strengthen the core values of the community. They combine familiarity and new and exciting activities to keep the community members engaged. And, through their activities they develop a rhythm for the community.

Communities that sustain also focus on issues and practices that are important to its members. They have a well-respected community leader who acts as facilitator or "community gardener." The gardener's role includes ensuring that people have opportunities to participate and are actively encouraged to participate.

Healthy communities of practice have a mixture of thought leaders and new learners. They help to build personal relationships among the members of the community and new members who join the community. The community provides a continuing forum for members to think together, to share information, and to pass on core knowledge.

We also know that communities of practice that arise from the interest and passion of their members have a greater chance of succeeding than those imposed by organizations or associations. Communities of practice resemble living organisms in that they have a natural life cycle. They emerge based on an identified need, they grow through the development and renewal of practice, and they mature and they fade away or pass into memory. One of the major distinctions between a naturally occurring community of practice and a designated or structured institutional community (i.e., project team, functional unit, administrative unit, committee, or task force) is that communities of practice live and die based on the interests, goals, and needs of the community members rather than by decisions made by individuals outside of the community. The community members, by their actions, determine when the community has fulfilled or outlived its purpose. To the extent that community members continue to derive value from a community of practice and find meaning in its existence, the community will continue to exist—not just over months and years but decades and centuries in some cases.

Healthy and sustainable communities of practice may have short or long life cycles. The life span of a community of practice is determined largely by the value that the community members derive from it. As long as the community offers value, as long as there are community members participating, and as long as practice continues to develop, the community will survive. A shorter life span does not necessarily mean the community failed. Neither does a longer life cycle denote success. The challenge for knowledge professionals, though, is that there are fewer examples of long-lived communities of practice than there are short-lived communities. This presents challenges for increasing our understanding of what works and what does not in this context.

The life cycle of a community of practice has been a primary source of interest among knowledge management researchers. Life cycle is closely aligned with community cultivation and community engagement.

CHARACTERISTICS OF COMMUNITIES OF PRACTICE

From the body of knowledge about communities of practice, we know that there are three dimensions of a community of practice which are important to its existence and survival:

- The domain in which they function
- The community and members
- The community's practice

A community's *domain* is defined by who participates and who is a member of the religious domain. Participation evolves over time through engagement, imagination, vision, and alignment—all continuous processes within the community of practice. The *community* dimension is defined not just by who belongs to the community but by the way the community works, the relationships it develops, and the shared experiences that become part of the way it works over time. Community includes how the members engage with one another, the sum of their participation and negotiated experiences, and the members' sense of belonging to the larger community. *Practice* includes what the community does, the meaning and value it creates, the actual practices that it develops and sustains, and how it grows and learns through the development of those practices. We will use these three dimensions to explore whether religious communities are communities of practice.

INTERPLAY BETWEEN RELIGIOUS COMMUNITIES AND KNOWLEDGE MANAGEMENT

The knowledge management literature makes reference to other historical forms of communities of practice (i.e., trade guilds of the middle ages, early scientific academies, etc.), but the most common form of communities of practice—religious communities—has been overlooked. There are two reasons for this. First, knowledge management professionals tend to focus on the creation and transfer of organizational knowledge—knowledge that pertains to the world of work. For this reason, we tend to focus on communities of practice that exist in the general work environment rather than the general social or spiritual environments. Second, we tend to think of religious communities as structured or institutional communities because of their doctrinal knowledge base and established religious and spiritual practices.

If we look more closely at religious communities, we find that they are anchored in many types of knowledge: doctrinal knowledge and scriptures, ritual knowledge and rites, individual spiritual beliefs, and religious teachings and religious community practice (i.e., living one's beliefs). We can view a "religious community" from many levels: the denominational level, the diocesan or administrative level, the church community, and the individual church member. Our focus in this chapter is on the church community level. At the individual church level, religious communities can exhibit all of the characteristics of communities of practice. In studying individual church communities, we can see how communities address issues of boundary and identity, how they develop practice in the form of religious teachings and ministries, and how they build and rejuvenate active congregations by enabling their individual members to engage in turning religious teachings in active practice. In the religious environment and at the church level, this practice takes the form of community activities and ministries. We tend to think of religious communities as long-standing and institutional with life cycles aligned with decisions made by denominational or diocesan leadership. However, at the church community level, we see the interplay between community characteristics, methods of cultivation, and life cycles, just as we do in other kinds of communities. Considered at this level and in this way, religious communities of practice provide a rich environment for knowledge professionals to increase their understanding of how communities of practice work.

Today, across the United States, economic and social factors are contributing to the loss of religious communities of practice (i.e., churches). The mainstream media reported on church closings that are the result of decisions from the church hierarchy. We can find examples of other church closings that resulted from a failure to rejuvenate the church community (i.e., membership). Our observations of church communities suggest that where churches behave like and adopt community of practice cultivation methods, those churches will survive and thrive. Although our observations are limited to a few case studies, only two of which are referenced here, they indicate that there is opportunity for religious communities to learn from knowledge professionals, as well as opportunities for knowledge professionals to learn from religious communities.

From the knowledge management perspective, long-lived communities of practice provide a rich environment for increasing our understanding of success and failure points. Communities of practice can emerge and disappear faster than academicians can find and study them. Although studying communities of practice with shorter life spans enables us to learn, these examples may tend to provide an incomplete perspective. Finding sustainable, regenerative communities of practice can help us to better understand

- Community processes of knowledge creation, transfer, mobilization, and capture

- The role of knowledge in building practice within the community and considering whether a lack of foundational knowledge is a potential point of failure
- The factors that contribute to long-term sustainability

RELIGIOUS COMMUNITIES OF PRACTICE—CHURCH COMMUNITIES

Let's begin by characterizing a religious community as a community of practice for knowledge professionals. According to Wenger, a community is defined by its domain, its community, and its practice. What defines the domain of a church community? What constitutes the community? And, what defines the nature of practice in this context?

Religious Domain

The religious domain is composed of its identity and its boundaries. Identity is defined by who participates and who is a member of the religious domain. Participation evolves over time through engagement, imagination, vision, and alignment—all continuous processes within the community of practice. In the religious domain, identity is all about creating bonds in which the community members become invested. Community members determine where they fit, who they are through commitment, development of trust, and perhaps most important of all for our religious communities, the development of shared histories. Identity is developed through engagement in common practices—in events of lived experiences.

Boundary is another aspect of the religious domain. Boundaries are identified by artifacts, connections, and relationships. Boundary artifacts may be real or abstract—they may be physical places or cultural traditions. Boundaries help religious communities to differentiate practices, to determine what is included or excluded from practice. Boundaries may be strongest within religious communities when they are defined around religious doctrine or religious rituals. However, boundaries may be weaker when we consider the translation of religious teachings to everyday life, to social and economic contexts.

Religious Community

According to Wenger, there are three aspects to the community: mutuality of engagement, negotiated repertoire, and accountability to the enterprise. By mutuality of engagement we mean the learned ways of engaging with others—the particular way in which the community goes about developing its practice. Engagement helps to define the nature of the community. In the religious communities we will discuss in our case studies, engagement is a critical success factor. In fact, it would appear that where the religious community actively engages its members in practicing or living its religious teachings, we find long-lived and sustainable religious communities.

Negotiated repertoire includes the development of shared experiences, language, artifacts, histories, and practices—built up over time as the community evolves. Negotiated repertoire is also a means of preserving and passing on these religious traditions and practices. Finally, accountability to the enterprise is achieved when community members take individual responsibility for the health and development of the community.

Religious Practice

Practice consists of three components: meaning, locality, and learning. In the context of religious communities, meaning is developed through participation, through mutual recognition, and through the shaping of a common experience over time, over generations, across and within individual religious communities (churches and parishes). As religious communities work together and workshop together, they develop a common purpose and meaning. Meaning becomes very strong when it grows over several generations, and in some cases over centuries. Locality in the religious community context is not limited to geographical proximity—it also suggests a "locality of interest," a commonality of religious beliefs, religious knowledge and teachings, and religious practices. Locality also includes local cultures—what the community does and how it grows and nurtures what it does. This includes the local adaptation of culture and the development of local norms. Learning and remembering and passing on common memories are essential to developing the power of practice. Community members become invested in other members of the community through this common learning activity. Artifacts and symbols are developed which perpetuate the practices with which they are associated. This is a key task in knowledge transfer—through learning, religious teachings and knowledge are passed from generation to generation.

Studying religious communities of practice raises a central question: What is the nature of religion and what is the nature of religious practice? There are many forms of religious practice and religious knowledge. According to the literature of sociology and religion,[35-40] religious practice and religious knowledge may include

- Religious doctrine as codified in a religion's sacred texts, its historical religious knowledge base such as the Bible, the Torah, the Koran, the Book of Mormon, and the interpretation of those sources in the religious writings of the scholars of the faith. Religious doctrine is generally developed by a select group of individuals in the religion, though its interpretation to practice involves a wider community.
- Religious rituals and shared beliefs as practiced in church religious ceremonies and observances.
- Religious teachings and sermons as prepared and delivered to the religious community by religious leaders.
- Everyday spiritual practice by individual members of the community.
- Church activities and ministries: the interpretation of church teachings and beliefs in everyday living which engage the members of the individual church on a daily basis.

Even though there are "communities" associated with each of these aspects of religious knowledge, church activities and ministries most closely resemble communities of practice from the knowledge professionals' perspective. This level of religious practice, often referred to in the literature as religious praxis, also appears to be the most influential for the sustainability of individual churches.

Two Case Studies: Church Communities Leveraging Knowledge Management Methods

We describe two vibrant religious communities of practice as case studies to illustrate our two points: (1) religious communities are valuable to the study of knowledge practices, and (2) knowledge management practices provide important methods for sustaining and rejuvenating religious communities.

CASE STUDY 1   Birth of a New Religious Community from Two Declining Parishes

Our first case study focuses on a parish in a large city in the Midwest. The parish in this first case study leveraged community of practice and knowledge management methods to develop a new and stronger religious community of practice through the blending of two well-established communities. In this case, the motivation to create a new religious community derived from a diocesan decision to close two historical parishes for economic reasons.

In the evolution of this new religious community, we can see community of practice cultivation methods at work. The new church community established a common domain from two distinct domains, built a joint community from the two existing communities, and expanded the community to include others who were not affiliated with either of the two original churches, and constructed new practices while preserving the memory of the old. This was not a trivial process, and it evolved over many years. The new church community faced several challenges. Both parishes were well established. Both churches had strong and independent identities, clear cultural boundaries, active though aging communities, and distinct practices. Both parishes had a common set of religious beliefs, religious doctrine, and rituals, but they had clearly evolved distinct church-level activities and practices over the past century and a half. The church activities and ministries reached deeply into the everyday life of the individual parishioners.

COMMUNITY IDENTITY

The first church was established in the first half of the nineteenth century with strong roots in Irish and German culture. The physical church that housed this community was a well-known historical landmark in the community. The physical church reflected elements of the two original cultures of the religious community. The community for this church was also geographically bound to the population by the diocese. The second church was established in the second half of the nineteenth century within the same city. In contrast to the first church, this church had no formal geographical boundaries. The identity of this community was well established in the Polish community.

In the early part of the twenty-first century, a new parish was formed with a new identity with expanded boundaries, new practices, renewed traditions, and an expanded and rejuvenated community. The new parish represents a spiritual and community building process that was facilitated by a spiritual leader whose methods resembled community of practice and knowledge management best practices.

Each parish developed a strong identity over more than a century. Each had a strong religious identity grounded in a system of personal faith and beliefs. Each had evolved a shared history in their new country as the community members both retained and adapted their traditions, cultures, and language. Each church had its own spiritual leader and its own historical place of worship. The second aspect of identity—boundaries—was strong in the two original parishes. Each parish developed explicit and long-standing boundaries reinforced by physical places of worship, religious schools, and in one case an explicit geographical boundary and in the other a strong boundary aligned with culturally defined communities. In order for the two distinct parishes to survive and thrive, a single new identity was needed. In order for a new identity to emerge, a boundary spanner was needed. In this example, the boundary spanner was a parish priest—a single spiritual leader with strong family ties to one of the original parishes. The parish priest built a core group of members, also boundary spanners, from the two original parishes. The priest provided an anchor and a common source of trust for the new emerging community.

A new identity with elements from the original two parishes was constructed through the construction of a new church. Elements of both former parishes were included to ensure that memories and traditions would be carried forward and kept alive. One of the former churches was closed and sold by the diocese, but elements of that church were used in the construction of the new building. The two communities leveraged the design and construction process to merge their separate physical and cultural identities into one new one.

COMMUNITY PRACTICE

As noted earlier, religious communities are complex entities with multiple levels of practice. One level of practice is that of religious denomination or diocesan. Practice at this level means a common doctrine, commonly held beliefs, and a calendar of religious observances and practices. A second level of practice is personal—an individual's interpretation of religious beliefs and his or her daily spiritual practices. A third level of practice is at the church or parish level—at the community level. At the church level we may find a commonly developed set of activities and ministries, local traditions, and rituals. Festivals, spaghetti dinners, church bazaars, church picnics, outreach activities, and ministries to the poor or to the isolated members of the larger community are examples of religious practice. The literature on religious knowledge, though, tends to focus primarily on rituals as spiritual practice. Through the development of this broader set of community practices, religious communities appear to develop, mature, sustain, and thrive. Individual churches also tend to have governance councils, or leadership councils. Church leadership is another key aspect of practice. In our case study, the two leadership councils were key to forging new practices. The new leadership provided core members of the new community.

Building a new and inclusive practice, though, was not a trivial task. The process used by the parish priest resembled how many community of practice today evolve practice. The first step was to develop a new parish directory with photos and information about parishioners. A second step was to develop a women's group to create a new parish cookbook, including recipes from both original parishes. The most significant effort, though, was the design and delivery of a community festival. The festival was a very interesting community activity—neither parish had previously undertaken such a large effort. It was a project that had a risk of failure, so there was a need for all community members to work together to ensure its success. All aspects of the festival were new, so in this way it provided community members with many opportunities to learn how to work together and to develop new practices. The festival provided a sense of accountability to the new parish, forged by individual parish members. New ways of doing things developed through the planning and delivery process created a new negotiated repertoire.

The festival was only one new practice among many. The parish priest realized the importance of blending religious practice and teachings and everyday life. In a religious community, practice often takes the form of ministries. The new parish established several new ministries, which according to the parish priest is the true focus of the church. New ministries included a social committee that provides planned activities that continually bring people in the parish together in a social setting, reinforce trust, and build personal relationships. A second ministry was an outreach to the poor and needy in the physical community surrounding the parish. This also included a food pantry. This ministry was one that grew considerably in membership and commitment within the new parish. A third ministry focused on hosting homeless families as part of an interfaith network. Additional ministries focused on Bible study, traditions of faith and faith formation, and adult education. Through these ministries and the initial blending and establishment of practice, the parish priest created a vibrant community of practice environment with extensive opportunities for participation.

COMMUNITY

In the case of religious communities from the nineteenth and early twentieth centuries, a religious community includes more than just the church. The concept of the community can extend beyond the church. Community can also include schools, clubs, family ties and relationships across generations, relationships with local businesses, and also political activities. Religious communities may, therefore, require more extensive networking and facilitation than other types of communities.

In our first case study, the parish priest fostered subcommunities that bridge all of these extended communities. He was actively involved in individual meetings and enabled new interpersonal relationships to develop. He also sought out and developed new leaders to move the new community forward. And, perhaps most importantly, the parish priest focused on building interpersonal relationships between members of all age groups.

A new vibrant community was forged from two separate and distinct parishes. The new community has a common culture, a common set of beliefs and practices, newly forged bonds, and new traditions. The members of the new community have a vested interest in its success. Although the original impetus for change came from outside of the community, the community brought about change and evolved with change at an individual activity level.

CASE STUDY 2   Birth of a New Religious Community to Address Community Needs

Our second case study of a church-level community of practice focuses on a community church in suburban Maryland. This community represents a different religious denomination—Baptist. It also represents a different context—the community developed around a strong set of ministries and an inclusive set of church practices that filled gaps in the social fabric of the larger community. The relationship of the community to higher levels of the denomination is not as strong as in the first case study. This church grew out of a need within the broader economic and social environment for a community where religious practices could develop and be realized.

COMMUNITY IDENTITY

In the case of the Baptist church, identity is clearly grounded in religious beliefs and teachings. This church is a relatively young church. It draws members from a wide geographical area that includes many other Baptist churches as well as other denominations. Church identity is observed in bumper stickers proclaiming affiliation with the church. These bumper stickers are seen miles away from the actual physical location of the church. Even though this area of Maryland has a 400 hundred year history, and several churches with strong historical roots, this church is not affiliated with any particular culture or any historical community. It includes members who are new to the area as well as those whose families have lived in the area for centuries. It crosses cultural and ethnic boundaries and also has a multiracial identity. Identity has clearly been forged anew by the community members.

Another interesting aspect of this church's identity is the fact that while the spiritual identity of the members is Baptist, the Baptist denominational affiliation is not an explicit identifier on any of the church literature, artifacts, or facilities. What one sees in terms of identifying factors are the individual ministries and practices of the church. One has to look deeper to see the denominational identity. It is clearly the church community's activities and ministries which define this church's domain.

COMMUNITY PRACTICE

What is most obvious about this church is its wide range of ministries—ministries that seem to touch all aspects of the lives of the community members. By ministry we do not mean "preaching" in the traditional sense but the translation of religious beliefs and teachings into everyday practice. This church supports a wide range of ministries—some focused on children, some on young adults, others on members with substance abuse challenges, and seniors. There are extensive early child care activities, after-school care, and summer vacation activities for children. There are extensive Bible study activities for adults and senior adults in particular. And there are weekly advertised meetings for substance abuse counseling. The church also supports many forms of celebrations, all openly and publicly advertised, to which the larger community is invited. And each ministry has at least one supportive subcommunity that enables individual parishioners to become engaged in and define that particular ministerial practice.

As far as can be determined in this superficial treatment of a very active community, the ministries of this church focus on its community members rather than the members of the external surrounding community. Practice focuses internally on the needs of the church community. The reach of the ministries—the religious practice—expands through the expansion of the community. A key method for community expansion is focus on young members of the church. In order to become part of the practice or derive value from the church's practice, you must become a member of that community.

COMMUNITY

The community members of this church are much younger than most other religious communities in the area. The community appears to revolve around young families and the supporting relationships among families. This may be a result of the surrounding economic and social environment, and another indication of the important of translating religious teachings into religious practice on a daily basis. Children are a very important aspect of this religious community. Unlike the first case, this church has no formal affiliation with school, business, or political activities. The community is solidly grounded in practices developed within the church community. Community members are drawn into the church, rather than the church extending outward. This continual draw into the church, of new members and families, helps to sustain and rejuvenate the community.

Another aspect of community which is different in this case is the nature of the community facilitation and leadership. Unlike the first church that revolved around a single spiritual leader, this church has several individuals who play leadership roles and have strong facilitation and communication skills.

Observations from the Case Studies

Our case studies are exploratory rather than definitive. The case studies suggest that there are opportunities for church-level communities to learn from knowledge professionals, and opportunities for knowledge professionals to learn from church communities. What can church communities learn from knowledge management? Our brief explorations into this area suggest that

- Church-level communities are critical to the survival of religious denominations: This is the level of practice that continues to keep the denomination alive.
- Community leaders who understand communities of practice can be successful at creating new and sustaining existing religious communities. Where there is a religious leader who knowingly or unknowingly guides the religious community as a facilitator practices knowledge management methods, the religious communities appear to sustain and regenerate and the communities replenish themselves.
- Domain is something defined through the building of the community—church community identity and boundaries need to be defined by the parish—rather than at a higher level as is commonly understood.
- Religious communities can be effective communities of practice when the community has an active repertoire of ministries. Ministries and church community building activities are important, regardless of whether they focus inward or outward.
- The ministries may be inward or outward looking. The key would appear to be the translation of religious belief and teaching into religious practice. Those communities of practice that focus on translating religious teachings into practice appear to be the ones that sustain and rejuvenate.
- The greater the mutual engagement within the church community, the stronger the community will be, and the longer it will sustain.
- It is important to enable subcommunities to form around ministries (children and youth, social problems, economic challenges, seniors, etc.) as parishioners become actively engaged through these subcommunities.

What can knowledge professionals learn from religious communities of practice? Again, drawing from these limited experiences, we suggest that

- There are many levels of community at play here. We need to look beyond the obvious form of community to see the true level of engagement and practice within the religious and spiritual community. Churches within our own communities are important sources of knowledge about how communities of practice work.
- The main lesson the researcher drew from these case studies is the importance of three elements all acting together: a set of common beliefs, shared repertoire, and mutual engagement. It is also important that church communities be allowed to define these three elements in today's complex social and economic environment.
- We understand that domain is clearly negotiated but that it must be negotiated by the community members, not handed down from a higher community or authority.
- Even though it may seem like an obvious conclusion, a church-level community is essential to the survival of the denomination. Where the community is not rejuvenating itself through practice that touches the lives of its members, the community will not survive.

REFERENCES

1. Etienne Wenger and Jean Lave. *Communities of Practice: Learning, Meaning, and Identity—Learning in Doing: Social, Cognitive, and Computational Perspectives.* Cambridge University Press, Cambridge, UK, 1999.
2. Etienne Wenger and Jean Lave. *Situated Learning: Legitimate Peripheral Participation—Learning in Doing: Social, Cognitive, and Computational Perspectives.* Cambridge University Press, Cambridge, UK, 1991.
3. Etienne Wenger, Richard McDermott, and William M. Snyder. *Cultivating Communities of Practice—A Guide to Managing Knowledge.* Harvard University Press, Cambridge, MA, 2002.
4. John Seely Brown and Paul Duguid. *The Social Life of Information.* Harvard Business Press, Cambridge, MA, 2000.
5. Etienne Wenger, Richard McDermott, and William M. Snyder. *Cultivating Communities of Practice—A Guide to Managing Knowledge.* Harvard University Press, Cambridge, MA, 2002.
6. Hubert Saint Onge and Debra Wallace. *Leveraging Communities of Practice for Strategic Advantage.* Butterworth Heineman, Oxford, 2002.
7. Nancy Dixon, Nate Allen, Tony Burgess, Pete Kilner, and Steve Schweitzer. *Company Command Unleashing the Power of the Army Profession.* Center for the Advancement of Leader Development and Organizational Learning, West Point, NY, 2005.
8. Peter Kilner. Transforming Army Learning through Communities of Practice. *Military Review*, 82(3, May/June), 2002.

9. Guillermo Palos. *Communities of Practice—Towards Leveraging Knowledge in the Military* (http://www.dtic.mil/cgi-bin/GetTADoc?AD=AD47443).

10. Eva Törnqvist, Johan Sigholm, and Simin Nadjm-Tehrani. Hastily formed networks for disaster response: Technical heterogeneity and virtual pockets of local order, in J. Landgren and S. Jul, eds., *Proceedings of the Sixth International ISCRAM Conference,* Gothenburg, Sweden, May 2009.

11. Peter J. Denning. Hastily formed networks. *Communications of the ACM,* 49(4, April), 15–20, 2006.

12. Christina Bollin. *Community-Based Disaster Risk Management Approach.* Deutsche Gesellschaft für Technische Zusammenarbeit, Eschborn, Germany, 2003.

13. Gita Swamy. Knowledge communities and the tsunami response: Experience from the Crisis Prevention and Recovery Community of the UNDP. *KM4D Journal,* 1(1), 57–61, 2005.

14. David Seaburn, Alan Lorenz, William Gunn, Jr., Barbara Gawinski, and Larry Mauksch. *Models of Collaboration: A Guide for Mental Health Professionals Working with Health Care Practitioners.* Basic Books, New York, 1996.

15. Noriko Hara and Khe Foon Hew. A case study of a longstanding online community of practice involving critical care and advanced practice nurses, in *Proceedings of the Thirty-Ninth Hawaii International Conference on System Sciences.* IEEE Press, 2006.

16. Nicola Andrew, D. Tolson, and D. Ferguson. Building on Wenger: Communities of practice in nursing. *Nurse Education Today,* 28(2), 246–252, 2008.

17. Nicola Andrew, Dorothy Ferguson, George Wilkie, Terry Corcoran, and Liz Sipson. *Developing Professional Identity in Nursing Academics: The Role of Communities of Practice.* Elsevier, New York, 2009.

18. Melanie A. Barwick, Julia Peters, and Katherine Boydell. Getting to uptake: Do communities of practice support the implementation of evidence-based practice? *Journal of Canadian Academy of Child and Adolescent Psychiatry,* 18(1), 16–29, 1999 (http://www.pubmedcentral.nih.gov/articlerender.fcgi?artid=2651208).

19. *Communities and Schools Promoting Health. Communities of Practice in School Health Promotion* (http://www.safehealthyschools.org/communitiesofpracticeintro.htm).

20. Vanessa Fuhrmans. Replicating Cleveland Clinic's Success Poses Major Challenges. *Wall Street Journal,* July 23, 2009 (http://online.wsj.com/article/SB124831191487074451.html).

21. Esther Green, Denise Bryant, Lisa Bitonti, and Debra Bakker. Communities of practice: Working in new ways to advance oncology nursing, presented at the *Fifteenth International Conference on Cancer Nursing,* Singapore, August 2008.

22. Helen Prytherch, Claudia Kessler, and Sandra Bernasconi. *2007 Evaluation of the SDC SOSA HIV/AIDS Community of Practice,* Swiss Centre for International Health, Basel, Switzerland, 2007.

23. http://www.sdc-health.ch/priorities_in_health/communicable_diseases/hiv_aids/2007_evaluation_of_the_sdc_sosa_hiv_aids_community_of_practice.

24. Alan T. Schussman. Uncertain certification: The problematic practice of wilderness medicine. American Sociological Association Meetings, August 2002.

25. Mark S. Schlager, Judith Fusco, and Patricia Schank. Evolution of an on-line education community of practice, in K. A. Renninger and W. Shumar (eds.), *Building Virtual Communities: Learning and Change in Cyberspace.* Cambridge University Press, Cambridge, UK, (pp. 129–158), 2002.

26. Joan Kang Shin and Beverly Bickel. Distributing teaching presence: Engaging teachers of English to young learners in an international virtual community of inquiry, in C. Kimble, P. Hildreth, and I. Bourdon (eds.), *Communities of Practice—Creating Learning Environments for Educators* (Vol. 2, pp. 149–178). Information Age Publishing, Charlotte, NC, 2007.

27. Eric M. Meyers, Lisa P. Nathan, and Matthew L. Saxton. Teacher-librarian communities: Changing practices in changing schools, in C. Kimble, P. Hildreth, and I. Bourdon (eds.), *Communities of Practice—Creating Learning Environments for Educators* (Vol. 2, pp. 199–222). Information Age Publishing, Charlotte, NC, 2007.

28. Wesley Shumar and Johann Sarmiento. Communities of practice at the math forum: Supporting teachers as professionals, in C. Kimble, P. Hildreth, and I. Bourdon (eds.), *Communities of Practice—Creating Learning Environments for Educators* (Vol. 2, pp. 223–240). Information Age Publishing, Charlotte, NC, 2007.

29. Chris Blackmore, Enabling duality in teaching and learning environmental decision making. A role for communities of practice?, in C. Kimble, P. Hildreth, and I. Bourdon (eds.), *Communities of Practice—Creating Learning Environments for Educators* (Vol. 2, pp. 309–326). Information Age Publishing, Charlotte, NC, 2007.

30. Dana Walker. The free knitting pattern collection: Connecting to the past—Shaping the future, *Social Systems and Collections*, December 2002.

31. Rhiannon Gainor. Leisure information behaviours in hobby quilting sites, *CAIS/ACSI 2009 Mapping the 21st Century Information Landscape: Borders, Bridges, and Byways* (The 37th Annual Conference of the Canadian Association for Information Science), Carleton University, Ottawa, May 28–30, 2009.

32. Ismael Peña-López. From social networks to virtual communities of practice. Beyond e-inclusion through digital literacy (I): The case of the crafting community (http://ictlogy.net/20081027-from-social-networks-to-virtual-communities-of-practice-beyond-e-inclusion-through-digital-literacy-i-the-case-of-the-crafting-community/).

33. Anne Ellen Geller. *Everyday Writing Center: A Community of Practice*. Utah State University Press, Logan, UT, 2006.

34. Edward Castronova. *Synthetic Worlds: The Business and Culture of Online Games*. University of Chicago Press, IL, 2006.

35. Peter L. Berger. *The Social Reality of Religion*. Middlesex, England, 1973.

36. Philippe Borgeaud. *La Memoire des religions. Religions en perspective*. Geneva, 1988.

37. Emile Durkheim, Mark S. Cladis, and Carol Cosman, *The Elementary Forms of Religious Life*. Oxford University Press, New York, 2008.

38. Leo F. Fay. Differential anomic responses in a religious community. *Sociological Analysis*, 39(1), 62–76, 1978.

39. Lester R. Kurtz. *Gods in the Global Village: The World's Religions in Sociological Perspective*. Sage, Thousand Oaks, CA, 2007.

40. Rick van Lier. New emerging religious communities in the Catholic Church of Quebec. *Colloqium: The Consecrated Life in Contemporary Canada*. Faculty of Religious Studies, McGill University, Montreal, Monday, May 11, 2009.

4

Cross-Cultural Technology-Mediated Collaboration: Case Study of Oxfam Quebec and Peru

Kimiz Dalkir

INTRODUCTION

Collaboration in our increasingly global work environment more than ever means collaborating with partners that are spread across the world. We need effective collaborative tools and effective collaborative strategies to be able to integrate cultural contexts that are often quite disparate. Technology-mediated collaboration already creates a significant challenge as it never quite approximates face-to-face interactions. Working at a distance in different countries only adds further complexity.

This case study describes a collaborative project that not only involved technology, due to the two groups being located in two different continents, but also differing attitudes and expectations of the role played by collaborative technologies, due to different cultural contexts. A knowledge management research team from McGill University, Montreal, Quebec, Canada, partnered with Oxfam, Ottawa, Ontario, Canada, and one of its partner units to collaborate on a knowledge transfer project between two countries, Canada and Peru.

The objective was to transfer best practices in youth empowerment and engagement ("mobilization") from a Peruvian organization to a Canadian-based one. The Peruvian organization had over 27 years of success in engaging youth to become actively involved in society through a series of cultural festivals. In Canada, on the other hand, most similar initiatives had not succeeded as well. The expectation was that the critical success factors identified in the Peruvian context could then be transferred to the Canadian context.

Initially, a series of telephone and e-mail-based interactions took place in order for both teams to collaborate on the design, development, and implementation of a Web site. This Web site was intended to be a collaborative medium where multimedia content would be authored by youth from both countries.

The McGill team conducted a postmortem exercise on work done prior to their joining the project. The researchers then worked with both the Peruvian and Canadian teams to develop collaborative multimedia content for a shared Web site.

BACKGROUND ON THE ORGANIZATION

McGill University is a research-intensive university based in Montreal, Quebec, Canada. The knowledge management (KM) research team consisted of researchers and graduate students from the School of Information Studies. Oxfam International was formed in 1995 by a group of independent nongovernmental organizations with the goal of reducing poverty and injustice. The name "Oxfam" was created in 1942. It is the shortened name for the Oxford Committee for Famine Relief that was used in telegraph communications.*

Oxfam is known for its work in providing emergency relief, as witnessed in recent events such as the Haitian earthquake. The organization also does extensive work in long-term projects to promote development in such areas as health, education, fair trade, gender equality, and alternative sources of food.

In addition to providing life-saving assistance to people in the context of natural disasters or man-made conflicts, Oxfam also works with international partners and communities on long-term programs to combat poverty and injustice. Oxfam is part of a global movement for change to raise public awareness of the causes of poverty and encourage ordinary people to take action for a fairer world. Oxfam presses decision-makers to change policies and practices that reinforce poverty and injustice in its partners in developing countries.

---

* From the Oxfam International Web site (http://oxfam.org).

There are 15 members of the Oxfam family, forming an international confederation governed by a secretariat that has partners in 98 countries around the world. McGill University partnered with Oxfam Quebec to work with one of its Peruvian partners, Vichama.

History of Oxfam Quebec in Peru*

Club 2/3 was, originally, a separate entity devoted to bringing together youth from North and South America in order to mobilize them as engaged citizens working toward a viable and equitable world. Mobilization refers to a mixture of awareness, autonomy, and intention to act on the part of the younger generation, who are all too often given to apathy rather than action.

In 2005, Club 2/3 joined with Oxfam Quebec. Club 2/3 has been working in Peru since 1981 and partnered on over 250 collaborative projects across the country. Today, Oxfam Quebec, together with Club 2/3, focuses its activities on access to clean water and basic education. They also provide support in other areas including technical and professional training for youth, improving living conditions for street youth, and developing collective entrepreneurship, particularly the cooperative model.

The McGill research team and Oxfam Quebec selected a specific project with a specific Peruvian partner as a pilot project to experiment with a number of collaborative channels and a number of different knowledge capture and transfer channels. Vichama is the Centro de arte y cultura, a theatre group from Villa El Salvador, Lima, Peru, that works worldwide, raising awareness of issues involving social responsibility and involvement.

Vichama is situated in Villa El Salvador, an urban center of about 400,000 inhabitants with a very strong sense of community. Members form a strong traditional social entity where everyone participates in development projects such as building houses, irrigation, parks, and so forth. The region has serious problems in that 30% of the children are malnourished, 54% are living in poverty, and 37% live in extreme poverty.

Vichama was the first nonprofit organization (NPO) to be based in this city. It is made up of a community of creative individuals dedicated to the democratization of art and culture. Their mantra is that art should be accessible to all. They provide a living theatre where people can interact with history, meditate around art objects, and otherwise actively engage

---

* From the Oxfam Quebec Web site: (http://oxfam.qc.ca).

with creative works. The Vichama community is vehemently opposed to art being only for the privileged. Since 1993, they involved youth in "weaving a social fabric out of art" by holding an annual fair where there are theatre workshops, community radio, film festivals, creation of murals, and other types of cultural exchanges.

In this way, Vichama has been quite successful in mobilizing youth over a significant number of years. Close to 50,000 participants attended the annual fairs since their inception in 2004. A number of activities have been filmed over the years, but most of the exchanges during the fair are verbal which means most of the knowledge exchanged remains in a tacit state. A substantial amount of social capital was created and maintained through Vichama's connections with a network of schools, teachers, parents, community leaders, state institutions, local government, the Minister of Education, and the National Institute of Culture. The context of these social interactions is that everyone is equal and that everyone has a voice. Vichama has had a long tradition of openness, transparency, mutual respect, and a fairly comfortable exchange of best practices and lessons learned. Learning from the past (organizational learning) is highly encouraged and has by now become an institutionalized practice.

Description of the Case Study

The pilot project on youth mobilization with Oxfam Quebec and Vichama was designed to provide a testbed to better understand the critical success factors behind Vichama's success in mobilizing youth. There were three separate entities involved: the Oxfam Quebec team working with Club 2/3 on youth mobilization, the Peruvian Vichama team working on youth mobilization, and the McGill research team studying their collaboration.

The McGill research team went to Peru during the annual Solidarity Cultural Forum in order to interview participants and to produce storytelling videos that captured and documented their best practices. These video stories would then be used to transfer the mostly tacit know-how gained from over 20 years' experience in mobilizing youth to the sister organization in Quebec (Club 2/3). A similar study of what worked and what could be improved was conducted with Club 2/3 members in Quebec.

Finally, a postmortem questionnaire was administered to both the Quebec and Peruvian teams in order to better understand the critical success factors involved as well as to better identify the similarities and differences in the two cultural contexts. It was important not only to understand why Vichama had been so successful but to also identify critical differences between the Peruvian and Canadian contexts in order to assess whether the transfer of best practices would be feasible. This type of comparative analysis is essential in the successful adaption of best practices by a different organization (Szulanski, 1996). The notion of knowledge "stickiness" was proposed by Szulanski (1996) to partially explain why knowledge transfer did not always succeed, as some knowledge remained "stuck" to its point of origin. Davenport and Prusak (1998) introduced a number of cultural factors that could make knowledge transfer difficult, such as lack of trust; different cultures, vocabularies, and frames of reference; lack of time and meeting places; a narrow idea of productive work; status and rewards accruing to knowledge "owners"; "not-invented-here" syndrome; and intolerance of mistakes or need for help. Above all else, they emphasize the importance of trust and common ground in facilitating knowledge transfer: "The closer people are to the culture of the knowledge being transferred, the easier it is to share and exchange" (p. 100).

At the end of the pilot project, an additional postmortem was held in order to assess how well the two teams had collaborated and whether the technologies they used were adequate and appropriate. This postmortem followed the exchange of informal assessments that seemed to point to a general sense that the collaboration had not gone as smoothly as it should have. There was also an intuitive sense that "cultural differences" were to blame. The postmortem session was videotaped and a facilitator was used to guide the session. Table 4.1 outlines the general protocol that was used.

The McGill team conducted this postmortem to better understand what worked well and what could have been improved with respect to how the two teams collaborated with one another.

Key Milestones

The Quebec and Peru teams starting working together in January. They held a series of phone calls to kickoff off the project. The Quebec team hired a Web designer for 3 months to work on the Web site where the best practices from Vichama would be documented so that Club 2/3 members could benefit from them. In February, the McGill research team went to Peru and documented some of the best practices in the form of video stories. This was done through interviews with individual participants and a Skype session with some of the original team members. The research teams' video stories were added to this Web site.

TABLE 4.1

Postmortem Protocol

Stage	Description	Notes
Introduction	Round table introductions; objectives of the postmortem; expectations of participants; how the results will be used	Explain the reason for videotaping is to do further analysis of the content later.
Description	Please characterize this project in terms of others you have worked on—length, complexity, partnering, and so forth	Typical or atypical project for the participants?
Best practices	Can you provide examples of what went as expected (routine)? Better than expected?	Identify best practices and any innovations.
Lessons learned	Can you provide some examples of what could have been done better? What would you change if you had to do this again? Why?	What do they feel could be some of the root causes of the problems they identify? Does everyone agree?
Feedback	Is this postmortem something you think you could do on your own for future projects?	Tell participants they will receive a summary of the session which they will be asked to validate.

Through the interviews, participants noted that after about 2 months, both teams began to realize there was a problem. The Quebec team felt the Web site development was proceeding at a snail's pace while the Peruvian team felt that the Web site was not at all representative of the type of content they wanted on there. After some analysis, it became apparent that the project was mostly perceived as a Web site development project by the Quebec team and given they had resources for only a limited amount of time, they were quite focused on meeting deadlines. The Quebec team felt they had a March deadline to meet. The Peruvian team felt that they were not on the same page, that they had not agreed on the Web site objectives, and that they were not being consulted as the Web site was being produced. The Peruvian team perceived this as a year-long partnership, with the physical Web site forming only a small part of the overall "deliverables" to be produced together.

The collaboration timeline began with mostly telephone calls and teleconferences in the first month, which then transformed into mostly e-mails during the next phases. There was never a face-to-face meeting of all participants. This was less problematic for the Quebec team, but the

Peruvian team members felt they did know who they were working with, they did not understand why the website design and content was being decided upon without their input and they did not have the same sense of urgency or the need to meet a deadline.

Although there were some linguistic challenges, native Spanish speakers were available to participate in most of the exchanges. However, there was definitely never a meeting of all the stakeholders involved through any technology medium. There was no project kickoff, although the Quebec team felt that the initial series of phone calls had served as a project kick-off. Instead, there was an assumption that some people would be present at some of the meetings and that any non-Spanish speakers would pick up the missing bits from their Spanish-speaking colleagues.

The cultural differences proved to be much more of a barrier to collaboration than language alone. The Vichama participants felt strongly that a community-based approach is key to development and progress. Their expectations of the Quebec team were to work in a highly collaborative manner. What emerged was that the two teams had a very different definition of "collaboration": the Quebec team felt that technology-mediated linkages meant that of course they were collaborating while the Vichama participants viewed this as communicating and connecting but not necessarily collaborating. They expected a higher level of active engagement, where everyone had an equal voice and where everyone was heard. They were less comfortable with the practice of "prototyping" a Web site—namely, putting something up and then having others critique it. These differences in expectations ultimately led to a breakdown in communication and the stalling of the Web site project.

COLLABORATION TOOLS AND PRINCIPLES APPLIED

A number of technology-mediated interactions have been studied to date, involving in roughly chronological order, teleconferencing, videoconferencing, Web conferencing, and Internet video phoning (e.g., Skype). The same research findings continue to apply to each and every technology: they never quite approximate the same conditions as face-to-face interactions (Fobler, 2008; McFadden and Price, 2007; Robinson, 2009). Technology obstructs collaboration to some extent, but different types of collaborative tools have different advantages and disadvantages. One measure that can be used to compare and contrast the degree to which technology helps or hinders collaboration is to look at the degree of social presence they can provide.

Technology-mediated communication, connection (social media), and collaboration channels can be characterized with respect to the degree of social presence they can provide. Social presence can be defined as the degree to which individuals perceive that they are interacting with another human being when participating in technology-mediated interactions (Dalkir, 2011). Video, which provides visuals, offers a higher social presence than teleconferencing, which offers only audio. With video, participants can see differences in facial expressions and body language cues. During a teleconference, it is possible only to perceive differences in tone. E-mail often offers the least amount of social presence (albeit the use of emoticons serve to increase social presence somewhat). Thurlow's Social Presence Model (2004) that ranks different channels notes that letters have the lowest social presence while face-to-face communication has the highest. McGrath (1990) places e-mail lowest, followed by teleconferencing, desktop videoconferencing, and face-to-face meetings.

The higher the social presence the more effective the collaboration will be (Dalkir, 2009, 2011). When participants have met face-to-face (ideally) in an initial meeting, subsequent technology-mediated interactions are much more natural and therefore more effective. Ideally, all first meetings should be face-to-face or through a technology that at least allows visual display (e.g., videoconferencing, Web conferencing, Skype, etc.). Participants form social connections even within a fairly limited amount of interaction time. They are more comfortable with one another, and their subsequent collaboration, even if technology mediated without visuals, will be more positive (Hurst and Hunter, 2002).

Isaacs and Tang (1993) were among the first to note that the use of video in remote collaboration improved the ability to show understanding, forecast responses, give nonverbal information, enhance verbal descriptions, manage pauses, and express attitudes. This has been validated quite strongly since, and the same advantages are proffered by other visual interaction media such as Skype.

The two teams collaborated using primarily telephone, teleconference, e-mail, and Web site postings. It was only during the McGill research team facilitated postmortem session that Skype was used. The Quebec team favored e-mail and posting changes on the Web site for the team in Peru to review. The Peruvian team preferred technologies with higher social presence such as Skype. These different preferences became very apparent during the team postmortem session held via Skype. The Peruvian team took the time to "greet" everyone, virtually shaking hands and showing participants around the home at which they gathered. Family members were present and introduced, and there were snacks to be shared virtually. The Quebec team was quite surprised, while the Peruvian team felt they had finally "met" after the project had been completed.

It became quickly apparent that had Skype been used in an initial project kickoff meeting, where participants could have met and established some degree of social connection, trust, and comfort, then the project would have turned out to be a great deal more successful.

Wainfain and Davis (2004) surveyed the effects of different communication media on the effectiveness of virtual collaborations. The authors analyzed more than 40 years of research on the effects of different forms of media and concluded that

> All media change the context of the communication somewhat, generally reducing cues used to (1) regulate and understand conversation, (2) indicate participants' perspective, power and status, and (3) move the group toward agreement. (Wainfain and Davis, 2004, p. xiii)

For example, in technology-mediated interactions, participants tend to cooperate less than they would have had it been face-to-face. It is easier to disagree at a distance. The authors conclude by saying it is imperative that people know each other personally before relying upon virtual collaboration. If this is not possible due to the distances or costs involved, then time should be taken to "break the ice"—to let participants socialize a bit and take the time to reach a common understanding of what they are trying to achieve and how they will do so. They recommend making a conscious effort to do so and employing someone to facilitate this ice-breaking session. Other recommendations to help make social connections between the participants include sharing biographies and images, routinely stopping to check that everyone is on the same page, and summarizing often during all technology-mediated meetings.

As it was, the Quebec team had "fast-forwarded" to the collaboration stage, when some of the initial foundation connecting work had not been done. This led to a rapid divergence in project goals with the

Quebec team focusing on just getting the Web site done because the Peruvian team was not commenting too much, while on the other hand the Peruvian team felt no ownership over the Web site. Vichama resides upon such a strong foundation of solidarity and community that they felt their only option was to opt out of a collaboration style that was too foreign to them.

The McGill research team had the advantage of meeting participants face-to-face in their own country. Two of the three team members were native Spanish speakers, and they were able to quickly establish trust and a social connection with the Vichama members:

> We at Vichama feel at ease that the McGill researchers are Latinas that speak our language and understand our culture...we feel that we are sharing our knowledge with people who are like us...our peers.

The major challenges faced in this case study were thus due to differences in technology preference, multilingual requirements for communication, and differences in culture that created different expectations and different attitudes toward collaboration practices. A variety of communication channels were used with varying degrees of social presence, but on the whole they had low social presence. Preferences differed between the two teams with respect to what types of technology to use in order to collaborate with one another. The Peruvian team clearly preferred channels with greater social presence such as those that afforded visual interaction, whereas the Quebec team was more used to tools such as teleconferencing and e-mail. Language differences created the need for translation, sometimes for simultaneous translation.

Differences in culture were significant which meant that participants needed to establish a rapport before beginning any actual collaborative work. This was a clear expectation on the part of the South American team. Vichama has strong roots in a culture of solidarity. The idea that a community progresses only through the participation of its citizens for the benefit of the whole community is a 30-year-old tradition that is still being practiced today. The culture in Quebec, as in most of North America, is of course quite different. In fact, researchers such as Kochan and Pascarelli (2003) and Esparragoza and Bodek (2007) found in their studies on cultural differences in mentoring, that Europeans and North American tend to view communalism as being dependent (i.e., not being independent and decisive enough on your own and requiring an entourage to "act as a crutch").

LESSONS LEARNED

Start Off on the Right Foot

The first lesson learned was to ensure that there was a kickoff meeting (at a minimum) that makes use of a technology with high social presence in order to establish a strong enough rapport among all team members. This kickoff meeting should have had two major goals: first, to connect participants and break the ice, and second, to reach a mutual consensus on what is to be done as well as how it is to be accomplished. The two teams had to spend enough time before starting to work together as they needed to develop trust and the basis for a social connection. Social activities such as sharing a meal, meeting family members, and so forth, served to decrease anxiety. In the absence of a face-to-face meeting, the next best thing would have been a videoconference or Skype or any other medium offering at least a visual interface. An initial meeting would also serve to have everyone agree on the communication and collaboration tools to be used for subsequent meetings. The next series of meetings could be held via lower social presence technologies such as teleconference or e-mail as long as the initial meetings had served their purpose in creating a comfortable and fairly close bond (e.g., a social network).

In fact, social network analysis could be done in pre- to poststrategy to assess the degree to which bonding occurred (Dalkir, 2009). The type of network analysis can be used to track who connected with whom (and using which medium). With time, interactions should be more frequent, indicating that perceived social distance decreased and active collaboration increased.

Project goals must also be clearly communicated to all participants prior to beginning the project. The differing cultural contexts need to be taken into account when discussing project goals. Context is closely related to culture—both are social not individual. Fullan (2001) defines context as structure, framework, environment, situation, circumstances, and ambiance. Changing the context can have powerful effects on how well project goals will be achieved.

Leave Your Titles at the Door

A second lesson learned was that hierarchies and status may need to be set aside in order to enable effective collaboration. Horizontal relationships represent the core of a collaborative team, and they should be the norm. Managers need to take on the role of a facilitator rather than a decision-maker. As stated by the McGill research team,

> Teamwork is mainly about situational leadership, letting the person with the relevant core competency for a given situation take leadership.

The differing perceptions of the Web site design meant that the Quebec team viewed this as a standard project, with project managers and deadlines and all the other key parameters of a project. The Peruvian team viewed the matter as a long-term collaborative partnership that would yield many projects, including the Web site project. They were far more preoccupied with the content rather than the development of the Web site.

As a result, the meetings were really at cross-purposes, with the Quebec team trying to manage a short-term project while the Vichama team was trying to embark on a collaborative partnership.

You Need People Who Can "Go Native"

Cross-cultural projects, especially if they involve virtual collaboration, should have intercultural participants. Intercultural participants are those people who possess the cultural know-how to act and integrate themselves into different cultures. They are critical to the success of a cross-cultural collaboration, even more so if the collaboration must take place at a distance with technology to mediate the interactions. Their role is to open dialogue between two different cultures and ensure that a reciprocal dialogue is established. In this case study, a great deal of tacit knowledge was required for the McGill research team to be able to integrate themselves into both cultures—Quebec and Peruvian. The fact that they were trilingual (French, Spanish, English) and two of the three were Hispanic (but not Peruvian) allowed them to contextualize the needed collaboration in all three contexts: McGill University, Vichama in Peru, and Club 2/3 in Quebec.

O'Brien et al. (2007) found in their study on intercultural student–teacher interactions that

> Teachers need to know how to instruct students in intercultural rhetoric, that is, how to persuade people to understand the way in which others located in different global contexts perceive, analyze, and produce situated knowledge. (p. 2)

The authors found that in order for technology-mediated cross-cultural collaboration and intercultural understanding to occur, dedication of focus to the task at hand, simulated proximity to the communicators, and close transparency of the medium were necessary. This is true in best practice transfers and organizational learning (Szulanski, 1996, 2000).

Take the Time to Reflect

A fourth lesson learned was that just as time was needed before collaboration began, time was also needed at the end of the collaborative project. Time needs to be set aside for reflection and for project postmortems. By identifying errors and how things can be improved in the next cycle, the Vichama team feels that they can learn from any mistakes and improve continually. They are in fact creating an organizational memory, which to date has remained tacit (transmitted orally) but which has began to take on a more permanent form through the video stories on the Web site.

For the McGill research team, these lessons learned were also valid internally. It was essential that there was a great deal of collaboration before departing for Peru. Both Oxfam Quebec and McGill team members met extensively to plan the logistics of the trip as well as how to collect data while there. As a result, the team was able to "hit the ground running" upon their arrival. This subsequently became a best practice for our partners, much like a briefing session companies have before sending employees to a country for the first time.

Collaboration Means Everyone Participates

Finally, a large part of the problems experienced during the online collaboration were due to the fact that the Web site was being "authored" by only one team. Given the community-based set of values of the Vichama team, something like a wiki may have been a good social media tool to implement. This would allow collaborative discussions and even codevelopment of the Web site. The perception would definitely be more of a tool that linked everyone and a place where everyone's voice could be easily heard (Williams, 2007). Mack and Mehta (2005) found similar benefits in using blogs that allowed all partners to post. They found this allowed for a more informal and more spontaneous form of communication that allowed participants to be more creative, definitely a good fit for the Vichama team. The authors found expressions tended to be more "what we feel" and "what we think" than "what we did." Both wikis and blogs afford greater social presence than e-mails and teleconferences, so the use of newer social media may serve to make asynchronous virtual interactions more intuitive.

FUTURE RESEARCH

There are still only limited studies that looked at the "inner workings of a team using online technologies in the course of their work" (Pretto and Pocknee, 2008, p. 780). Most studies focused on the use of mixed media for online collaboration, that is, a mix of face-to-face and technology-mediated collaboration (Allen et al., 2005; Beekhuyzen et al, 2006; Larsen and McInerney, 2002). Pretto and Pocknee (2008) looked at the case of three different universities collaborating together, and although they took the time and gave serious consideration to the basic collaborative tools they would use, they still went through a trial-and-error period. "Even the simplest communication systems were a struggle" (p. 781), and they concluded that "our naivety in believing that the technology was ready and able to conquer the 'tyranny of distance'" (p. 783) was misplaced.

There are so many levels of challenges, beginning with incompatibility between technological platforms, to cultural differences, that the role of technology in virtual collaboration teams cannot be taken for granted. Collaborative tools can, however, add considerable value to online project collaborations as long as they are supplemented by face-to-face interactions (Larsen and McInerney, 2002). The nature of this blended approach requires additional research in order to find the optimal balance.

The role of social presence also remains largely unexplored: Can high social presence tools such as Skype approximate face-to-face interactions enough to boost the effectiveness of online collaboration? Or is there something fundamentally different in face-to-face encounters which is a critical success factor for virtual collaboration? Finally, what role is or can be played by social media in bridging the technological issues and perhaps even the cultural ones? Finally, more longitudinal studies are needed in this area. Most studies look at interactions during the course of projects with a limited life span. The sustainability of collaboration is a further challenge that needs to be better analyzed. What are the critical success factors needed to ensure that a virtual collaborative effort remains effective for the duration of a project (in particular projects of a long duration)? The Vichama participants encapsulated their own lesson learned with respect to the sustainability of knowledge exchange during collaboration as follows:

> To form a group, is a start; to stay together is progress; to work together is to succeed.

ACKNOWLEDGMENTS

The author would like to thank members of the McGill research team, in particular, Evelyne Mondou (Project Manager), Rita Benitez, Nathalie Blanchard, Sarah Burns, Eric Erickson, Margaret Ferguson, Maimi Niina, Andrea Puhl, Diana Tabatabai, and Erica Wiseman, as well as our partners in Oxfam Quebec.

REFERENCES

Allen, R., Becerik, B., Pollalis, S., and Schwegler, B. (2005). Promise and barriers to technology enabled and open project team collaboration. *Journal of Professional Issues in Engineering and Practice*, 131(4): 301–311.

Beekhuyzen, J., Cabraal, A., Singh, S., and Von Hellens, L. (2006). Confession of a virtual team. In *Proceedings of the Third Annual QualIT Conference*, Brisbane, Australia. Available online at: http://www.cit.gu.edu.au/conferences/qualit2006/proceedings/

Dalkir, K. (2009). Computer-mediated knowledge sharing. In *E-Collaboration: Concepts, Methodologies, Tools, and Applications*, Vol. 1, N. Kock (Ed.). Hershey, PA: Information Science Reference, pp. 48–66.

Dalkir, K. (2011). Measuring the impact of social media: Connection, communication, and collaboration. In *Social Knowledge: Using Social Media to Know What You Know*. J.P. Girard and J.L. Girard (Eds.). Hershey, PA: IGI Global, pp. 24–36.

Davenport, T.H., and Prusak, L. (1998). *Working Knowledge: How Organizations Manage What They Know*. Boston, MA: Harvard Business School Press.

Esparragoza, I., and Bodek, M. (2007). Using interactive web conferencing for international collaboration with institutions in Latin America and the Caribbean. In *Proceedings of the Fifth LACCEI International Latin American and Caribbean Conference for Engineering and Technology (LaCCEI 2007)*. May 29 to June 1, Tampico, Mexico. http://www.Laccei.org

Fobler, F. (2008). *A Practice Theoretical Analysis of Real Time Collaboration Technologies: Skype and Sametime in Software Development Projects*. Göttingen, Germany: Cuvillier.

Fullan, M. (2001). *Leading in a Culture of Change*. San Francisco: Jossey-Bass.

Hunter, D. (2007). Online Facilitation. In *The Art of Facilitation* (Chap. 6). Auckland, New Zealand: Random House NZ, pp. 175–180.

Hurst, D., and Hunter, J. (2002). Developing team skills and accomplishing team projects online. In *Theory and Practice of Online Learning*, 2nd ed., Anderson, T. (Ed.), Athabasca, Alberta: Athabasca University.

Isaacs, E., and Tang, J. (1993). What video can and can't do for collaboration: A case study. In *Proceedings of MULTIMEDIA '93: Proceedings of the First ACM International Conference on Multimedia*. J. Garcia-Luna-Aceves and P. Rangan, Eds. ACM: New York. pp. 199–206.

Kochan, F., and Pascarelli, J. (2003). *Global Perspectives on Mentoring. Transferring Contexts, Communities, and Cultures*. Charlotte, NC: Information Age.

Larsen, K., and McInerney, C. (2002). Preparing to work in the virtual organization. *Information and Management*, 39: 445–456.

Mack, A., and Mehta, D. (2005). Accelerating collaboration with social tools. In *Proceedings of the Ethnographic Praxis in Industry Conference*, 2005(1): 146–152. Available online at: http://onlinelibrary.wiley.com/doi/10.1111/j.1559-8918.2005.tb00015.x/abstract (doi: 10.1111/j.1559-8918.2005.tb00015.x).

McFadden, A., and Price, B. (2007). SKYPE: A synchronous tool for computer-mediated collaboration. *International Forum of Teaching and Studies*, 3(2): 37–45.

McGrath, J. (1990). Time matters in groups. In *Intellectual Teamwork: Social and Technical Bases of Collaborative Work*. J. Galegher, R.E. Kraut, and C. Egido (Eds.). Hillsdale, NJ: Erlbaum, pp. 23–61.

O'Brien, A., Alfano, C., and Magnusson, E. (2007). Improving cross-cultural communication through collaborative technologies. *Pervasive Technology, Lecture Notes in Computer Science*, 4744/2007: 125–131.

Pretto, G., and Pocknee, C. (2008). Online project collaboration…We still have a long way to go. In *Proceedings of the Ascilite 2008 Melbourne Conference*. Available online at: http://www.ascilite.org.au/conferences/melbourne08/procs/index.htm

Robinson, J. (2009). Cultural exchange through Skype. In *Proceedings of the DigitalStream Conference*. California State University, Monterey Bay. Available online at: http://php.csumb.edu/wlc/ojs/index.php/ds/article/viewArticle/51

Szulanski, G. (1996). Exploring internal stickiness: Impediments to the transfer of best practice within the firm. *Strategic Management Journal*, 17: 27–43.

Szulanski, G. (2000). The process of knowledge transfer: A diachronic analysis of stickiness. *Organizational Behavior and Human Decision Processes*, 82(1): 9–27.

Thurlow, C., Engel, L., and Tomic, A. (2004). *Computer-Mediated Communication: Social Interaction and the Internet*. London: Sage.

Wainfain, L., and Davis, P. (2004). *Challenges in Virtual Collaboration. Videoconferencing, Audioconferencing, and Computer-Mediated Communications*. Santa Monica, CA: The Rand Corporation. Available online at: http://www.rand.org

Williams, A. (2007). *Wikinomics: How Mass Collaboration Changes Everything*. Toronto, Ontario, Canada: New Paradigm Learning.

5

Enabling Knowledge Exchange to Improve Health Outcomes through a Multipartner Global Health Program

Theresa C. Norton

INTRODUCTION

Global health programs have grown significantly in size and complexity over the past decade. Programs frequently involve multiple partners, such as national governments, multilateral agencies, corporations, nongovernmental organizations (NGOs), and private foundations (Esser, 2009). Large global health initiatives such as the President's Emergency Plan for AIDS Relief (PEPFAR) and President's Malaria Initiative (PMI) add to the complex environment with many partnerships among donors, implementing partners, and stakeholders. According to one report, there are now over 100 global partnerships in the health sector alone (Action for Global Health, 2011), while some donors are increasingly bundling disparate project activities in multisectoral "mega-programs" that yield more coordination challenges (ACVFA, 2007).

With increasingly complex global health programs, effective use of aid funds requires a rigorous approach to knowledge management—the systematic use of people, processes, and technology to capture and share "know-how." The flow of knowledge needs to occur within and among partner organizations, across geographic and language boundaries, so that program teams can learn from each other and function as a cohesive whole. Ultimately, global programs aim to scale up the adoption of high-impact health practices* for better health outcomes of populations. Meeting this goal requires that the right information gets to the right people at the right time—one of the basic tenets of knowledge management—in a way that will grab their attention, motivate them to act, and be compatible with the environment in which they act.

The case study presented below illustrates a mixed approach to knowledge management interventions used for a large-scale global health program. The interventions used provide for variations in technology access and networking preferences. A discussion of knowledge translation, social networking, and team collaboration concepts, drawn from literature, follows the case study to provide insight into the effectiveness of the approaches used.

BACKGROUND ON THE GLOBAL HEALTH PROGRAM

Since the 1960s, international health programs have explored the benefits of offering women family planning services in coordination with maternal health services in order to prevent unwanted pregnancies (Ringheim, 2011). Reports show that less than 5 percent of women in developing countries want to become pregnant soon after childbirth, yet more than one-third are pregnant again within 15 months (Ross and Winfrey, 2001). With the growth of programs to integrate family planning and maternal, newborn, and child health services, countries such as Thailand that made integrated services part of a national strategy realized an increase in contraceptive use, decrease in fertility, and strengthening of their economies (Ringheim, 2011).

Studies also show that family planning after childbirth (called postpartum family planning or PPFP[†]) has life-saving benefits as well. Evidence shows that averting unplanned/unwanted pregnancies, as well as using healthy birth spacing, saves lives of mothers and children (Campbell and Graham, 2006; Cleland et al., 2006). Yet despite the fact that the proven benefits of PPFP have been known for years, global health programs have been inconsistent in their approaches, resulting in a continuing gap in access and quality of healthcare services for mothers and their families (ACCESS-FP, 2006).

---

* High-impact practices, when scaled up and institutionalized, will maximize investments in comprehensive health strategies. (USAID, 2011). In other words, high-impact practices yield high returns in terms of health outcomes for the investment.
† For this paper, PPFP is defined as the first year after childbirth.

To respond to the need for stronger PPFP services worldwide, in 2005 the U.S. Agency for International Development (USAID) funded a 5-year program called Addressing Unmet Need for Family Planning in Maternal, Neonatal, and Child Health Programs (ACCESS-FP). The program, associated with USAID's flagship maternal and newborn health program called ACCESS, focused on reducing unmet need for family planning among postpartum women by strengthening maternal, neonatal, and child health service delivery programs (USAID, 2006). Program objectives included

- Testing alternative ways of delivering healthcare services to increase use of PPFP methods
- Improving use of the lactational amenorrhea method (LAM) of family planning* and transition to longer-term methods
- Promoting healthy spacing of births
- Looking for ways to integrate family planning services more effectively in maternal, neonatal, and child health programs

Jhpiego led implementation of the ACCESS-FP and ACCESS programs. An international, nonprofit health organization affiliated with Johns Hopkins University, Jhpiego is based in Baltimore, Maryland. Founded in 1974, Jhpiego has worked in more than 150 countries with health experts, governments, and community leaders to provide high-quality healthcare for their people. The organization develops strategies to help countries care for themselves by training competent healthcare workers, strengthening health systems, and improving delivery of care. In designing programs, Jhpiego focuses on practical, evidence-based interventions for low-resource settings.

Five implementation partners worked with Jhpiego on ACCESS-FP: Save the Children, the Futures Group, the Academy for Educational Development, the American College of Nurse-Midwives, and Interchurch Medical Assistance. Country stakeholders included ministries of health and healthcare facilities such as maternity hospitals and primary healthcare in 11 countries:

- Afghanistan
- Albania

---

* LAM involves use of breastfeeding to delay return to fertility, particularly in the first 6 months after childbirth.

- Bangladesh
- Burkina Faso
- Guinea
- Haiti
- India
- Kenya
- Nigeria
- Rwanda
- Tanzania

To continue the work of ACCESS, ACCESS-FP, and other maternal, newborn, and child health programs that were ending, USAID funded a large global health program called the Maternal and Child Health Integrated Program (MCHIP) in 2008. This $600 million program increased the number of partners and countries involved from the previous ACCESS-FP program—from five to eight partners, and from 11 to more than 30 countries. The partner organizations were selected as leading experts in their field with large-scale experience. Each partner organization takes the lead in developing projects around specific technical areas within the scope of MCHIP, which is somewhat broad, ranging from clinical areas such as maternal, newborn, and child health-care to nonclinical interventions such as social marketing. By scaling up evidence-based, high-impact maternal, newborn, and children health interventions, MCHIP aims to help reduce maternal and child mortality by 25% across 30 priority countries through field-based implementation and global leadership (MCHIP, 2011).

ROLE OF KNOWLEDGE MANAGEMENT IN THE ACCESS-FP AND MCHIP PROGRAMS

The requirements for ACCESS-FP and MCHIP include a key knowledge management objective to generate and share "program learning." In program learning, emphasis is placed on developing and sharing evidence in the field of what works well in meeting health goals (i.e., best practices), so that this knowledge can be documented, replicated, used, and scaled-up, and even improved upon in other settings. The implicit, multidirectional aspect of program learning in a complex program (among countries, among partners, etc.) poses knowledge exchange challenges from a number of standpoints, such as

- Engaging geographically dispersed program participants and potential beneficiaries which leads to challenges (e.g., time zone differences for real-time discussion, travel expenses, and language differences)
- Carrying out program activities in low-resource settings with unreliable Internet connectivity and limited access to scientific literature
- Encouraging dialogue among experts so that they reach consensus on evidence indicating best practices
- Obtaining documented success stories and lessons learned from busy program staff members and healthcare professionals in the field
- Effectively disseminating program learning so that it gets noticed and acted upon appropriately in an era of information overload

From a sociological standpoint, additional challenges influence the suitability of knowledge management or program learning interventions, such as

- Personal learning style
- Organizational culture
- Regional culture
- Political environment
- Social networking dynamics (linkages to other individuals and organizations) and preferences (face-to-face versus electronic contact)

ACCESS-FP and MCHIP approached these challenges by using a mix of knowledge management approaches that take into account personal, organizational, and environmental factors, including

- *Technical Consultation Meetings*: Face-to-face, technical* consultation meetings with experts and leaders in reproductive health and maternal, neonatal, and child health to review the state of the PPFP literature; share programmatic experiences, lessons learned, and tools; and recommend a research agenda (2006, 2008, 2009, 2010)

---

* The term *technical* refers to public health or medical information.

- *Community of Practice**: Formation of a PPFP community of practice (CoP) for continuing support and dialogue; communication through annual, face-to-face meetings, and online engagement (2006)
- *Online Collaboration and Forums*: Online collaboration of CoP members through a mega-CoP platform called the Implementing Best Practices (IBP) Initiative Knowledge Gateway (2006); includes a series of online global forums conducted via e-mail with guest expert "speakers" and message archives in the CoP collaboration area (2007 to 2011)
- *Synthesized Knowledge for Use*: Creation of an electronic toolkit to provide practical information for developing and implementing PPFP programs to assist policymakers, program managers, trainers, and service providers
- *E-Learning Course*: Development of a Web-based course for low-bandwidth Internet connection, part of USAID's Global Health eLearning Center, which offers more than 40 online courses for public health professionals working in the field, particularly those at USAID missions and USAID implementing partners

The programs use other knowledge management approaches as well, such as an online collaboration area for partners, partner events, conference presentations, and journal articles to promote awareness of evidence. This case study focuses on the approaches listed above. The following section provides more detail about each approach.

Building Consensus: Technical Consultation Meetings

Following an extensive literature review to determine the state of PPFP programming, ACCESS-FP and USAID convened the first in a series of technical consultation meetings with experts to discuss and analyze the findings of the review. Held in 2006 in Washington, DC, the first meeting brought together 40 experts from more than 23 global organizations. It had the following objectives:

- Develop guidance for providing PPFP services
- Identify gaps in knowledge and areas for future research
- Discuss opportunities and approaches to improving integration of maternal health and family planning services (ACCESS-FP, 2006)

---

* A "community of practice" is a group of people who share an interest in a common topic, such as family planning, and are engaged in activities related to that topic (e.g., as part of their job).

Participants represented a cross section of maternal, neonatal, and child health and family planning professionals.

The meeting format consisted of

- Presentations synthesizing the state of the PPFP literature
- Small-group work to identify gaps, recommendations, and opportunities for programs based on the literature review and the participants' experiences
- Group "report-outs" during a plenary session
- Call to action for participants (e.g., taking recommendations back to participants' programs)

The meeting format built a sense of team affiliation and encouraged consensus among participants. The initial presentations on state of the literature ensured that all participants started discussions with the same baseline knowledge of basic issues, an important step for team building and collaboration. With a consistent understanding of the issues, participants could then work together in small groups to analyze findings and evaluate possible actions. Finally, the groups were called to make recommendations on actions related to promoting PPFP. Participants were also strongly encouraged to continue the dialogue, by means of annual face-to-face meetings and as an online community.

The subsequent meetings held in 2008, 2009, and 2010 grew to include over 60 participants from global health organizations. These later meetings provided a forum for sharing programmatic strategies, state-of-the-art knowledge, and lessons learned (ACCESS-FP, 2010). Participants also shared pretested tools to support PPFP programming and prioritized programmatic topics for research and learning.

Perpetuating a Sense of Team Affiliation: The Postpartum Family Planning (PPFP) Community of Practice and Online Collaboration

Meeting participants were asked to perform an important follow-up action—use an online collaboration platform to continue the dialogue and form new linkages so that progress would continue beyond the 2006 meeting. Participants were asked to become members of the Implementing Best Practices (IBP) Initiative. Started by the World Health Organization and USAID, the IBP Initiative provides a forum for the global reproductive health community to share evidence-based practices that can be used in low-resource settings. Members receive invitations to face-to-face meetings and can sign up for a free account on the related IBP Knowledge Gateway (http://www.ibpinitiative.org/knowledge_gateway.html). The Gateway is an online platform designed for use by global health CoPs. Communities can request collaboration space on the Gateway where they can post announcements, create a community-specific digital library, and establish a discussion board. Use of the Gateway is free. Since its creation in 2004, membership in the Gateway has grown to 17,850 members from 215 countries and territories who participate in over 400 communities (O'Brien and Richey, 2010).

ACCESS-FP invited meeting participants to join the PPFP CoP housed on the IBP Knowledge Gateway. Members of the PPFP CoP receive notifications about events and activities related to PPFP and can participate in the online discussions. ACCESS-FP encouraged members to share their experiences and tools or other documents by posting them to the online collaboration area. As a result of continued dialogue through the online CoP space and the face-to-face meetings, the following four working groups were formed so that partners could address specific issues: (1) Lactational Amenorrhea Method and the Transition to Modern Contraceptive Methods, (2) Postpartum Intrauterine Contraceptive Devices, (3) Immunization and Family Planning Integration of Services, and (4) Infant and Young Child Nutrition and Family Planning Integration of Services.

Global PPFP Online Forums: Formal and Informal Learning

ACCESS-FP organized a series of global online forums that generate growing interest and involvement. From 2007 to 2011, nine forums offered "mini-lectures" on practical PPFP service delivery and programming in low-resource settings and engaged participants to share their experiences. Topics for each forum relate to the special interests and recommendations posed by participants at the technical consultation meetings.

The format of the forums consists of

- Two-week, moderated discussion via e-mail
- Daily digests
- Five to 10 global health experts per forum posting mini-lectures with discussion questions
- Web-based archives of messages

- Attachments for further reading
- A focus on practical experience, lessons learned, and evidence-based practices

The mini-lectures provide concise, formal learning opportunities for participants, while the online discussions with other CoP members provide informal learning opportunities.

The Knowledge4Health Project, led by the Johns Hopkins Bloomberg School of Public Health, Center for Communication Programs, and also funded by USAID, manages the online platform and provides usage statistics to the CoP facilitators. A coordinator from ACCESS-FP serves as facilitator for the PPFP online forums, managing memberships and posting daily digests. To help members feel a sense of connection, the facilitator notes in digests the countries that are actively participating in the discussions. Although Internet connectivity generally poses challenges in low-resources settings, a review of the digests and forum archives shows between 3 and 11 developing countries participating in each forum. (See Table 5.1.) This level of activity, along with anecdotal evidence, suggests that e-mail use is sufficiently low in bandwidth to enable online participation by developing countries that have some level of Internet connectivity.

Despite the bandwidth friendliness of e-mail communication, a review of the forum messages (many of which note the participant's organization) indicates that participation is largely from members who work for developed country organizations. Although these participants are based in developing countries, most are staff members from programs such as ACCESS-FP. Frontline healthcare providers in developing countries may not always be able to participate directly in the online activities; however, they can still benefit from and enrich the shared dialogue. One participant, a program officer in Nigeria, noted that he works in healthcare facilities that do not have Internet connectivity. He acts as the liaison between the forum participants and the frontline healthcare providers, passing along reports of field challenges to the forums and tapping into expert knowledge during the forum to gain insight into solutions. For example, a forum discussing women's misperceptions about return to fertility led the program officer to include the topic of return to fertility counseling in a service delivery supervision checklist (Figure 5.1) (Samaila Yusuf, personal communication, August 17, 2011). In other examples, during a forum on essential medicines, discussions

TABLE 5.1

Between 3 and 11 Developing Countries Post Messages on Each Global Forum

Forum Title	Date	Countries Posting Messages
Integrated Service Delivery of Immunization and Family Planning	July 2011	Bangladesh, Democratic Republic of Congo, Kenya, India, Nigeria, Tanzania, Vietnam
Maternal, Infant, and Young Child Nutrition and Family Planning Integration	February 2011	Afghanistan, India, Malawi
Online Forum on the "Guide to Developing Family Planning Messages for Women in the First Year Postpartum"	June 2010	India, Kenya, Nepal, Nigeria
Lactational Amenorrhea Method (LAM) and the Transition to Other Modern Methods	January to February 2010	Afghanistan, Albania, Bangladesh, Cameroon, Côte d'Ivoire, Ghana, Guinea, Madagascar, Mexico, Nigeria
Postpartum Intrauterine Contraceptive Devices (PPIUCD)	October 2009	Cameroon, Côte d'Ivoire, Ghana, India, Kenya, Pakistan, Uganda, Zambia
Strategies for Community-Based PPFP	March 2009	Bangladesh, Egypt, Ethiopia, Guinea, India, Kenya, Niger, Nigeria, Rwanda, Uganda, Zambia
PPFP Contraceptive Technology	September 2008	Cambodia, Democratic Republic of Congo, Georgia, Guatemala, Kenya, Malawi, Pakistan, Uganda
Key Messages for PPFP	April 2008	Bangladesh, Democratic Republic of Congo, Egypt, Guatemala India, Kenya, Mexico
Health Timing and Spacing of Pregnancy	November to December 2007	Democratic Republic of Congo, Egypt, India, Jordan, Kenya, Philippines

led a participant from India to use a new forecasting model to reduce stockouts (lack of supplies), and another participant reported plans to use the forum information to shape the Ethiopian government's assessment of reproductive health commodity procurement (O'Brien and Richey, 2010).

Of the CoP online activities, the online forums have the largest impact on membership. According to the Gateway records, over the course of nine forums, the PPFP CoP online membership grew from

Checklist Administration during Supportive Supervisory Visits to MCHIP Supported Facilities in Northern Nigeria

Name of Health Facility: _____ State: _____

LGA: _____ Date: _____

Completed by: _____ Signature: _____

Specific checklist	Tick if completed during the visit	If not completed during the visit, state reason/s	Comments
IUD Counseling & clinical skills			
FP counseling skills			
Postabortion care clinical skills			
PPFP Counseling skills			
Choice of methods for PPFP			
Counseling for PPFP clients—immediate postpartum and six weeks postpartum (exclusive breastfeeding, return to fertility, HTSP, etc.)			

FIGURE 5.1

A postpartum family planning (PPFP) community of practice (CoP) online forum discussion about return to fertility helps craft a service delivery checklist. (LGA stands for local government area; HTSP stands for Healthy Timing and Spacing of Pregnancy). (From Samaila Yusuf, personal communication, August 17, 2011.)

200 members in 35 countries in 2007 to 976 members in 79 countries in 2011. (See Figure 5.2.)

Postpartum Family Planning e-Toolkit: Synthesized Knowledge

As part of the effort to help program staff locate practical information to improve PPFP programs, ACCESS-FP used an electronic toolkit model developed by the Knowledge4Health program. The e-toolkit provides a

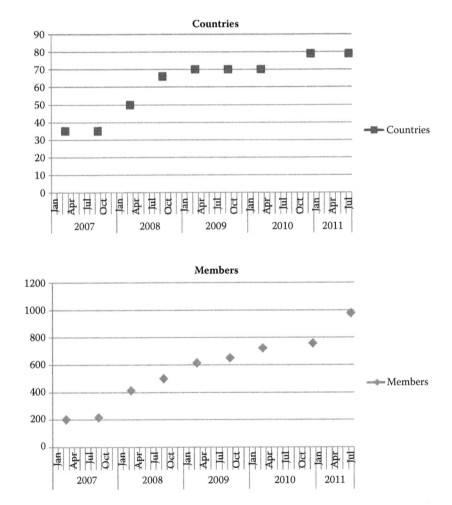

FIGURE 5.2

Postpartum family planning (PPFP) community of practice (CoP) online membership grew in number of members and countries over the course of nine online forums (2007 to 2011).

comprehensive collection of best practices and evidence-based tools and documents on PPFP to assist policymakers, program managers, trainers, and service providers. (See Figure 5.3.) The Web version of the toolkit is housed on Knowledge4Health's Web site at http://www.k4health. org/toolkits/ppfp. Knowledge4Health also distributes offline versions of e-toolkits on flash drives.

By using Knowledge4Health's existing e-toolkit platform, which was already funded by USAID, ACCESS-FP was able to make prudent use of USAID funding and avoid creating a duplicate platform. In addition, the e-toolkit skeleton is a template used to create 36 toolkits from 70 organizations on global health topics to date. Use of a replicated model benefits users of the other toolkits by providing a recognizable format.

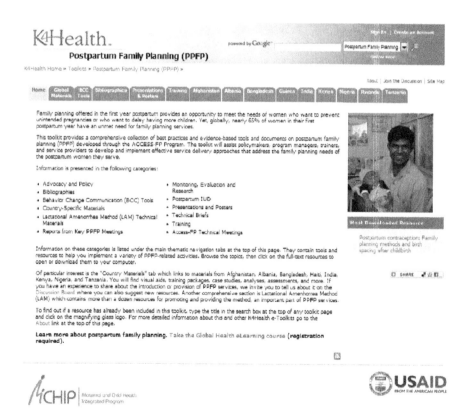

FIGURE 5.3

The postpartum family planning (PPFP) toolkit provides a comprehensive collection of best practices and evidence-based documents to support programs. (Used with permission from Knowledge4Health Program at Johns Hopkins Bloomberg School of Public Health, Center for Communication Programs, http://www.k4health.org/toolkits/ppfp.)

Information in the PPFP e-toolkit is presented in the following categories:

- Advocacy and Policy Bibliographies
- Behavior Change Communication (BCC) Tools
- Country-Specific Materials
- Lactational Amenorrhea Method (LAM) Technical Materials
- Reports from Key PPFP Meetings
- Monitoring, Evaluation, and Research
- Postpartum IUD
- Presentations and Posters
- Technical Briefs
- Training
- ACCESS-FP Technical Meetings

Country-specific materials from Afghanistan, Albania, Bangladesh, Guinea, India, Kenya, Nigeria, Rwanda, and Tanzania are included.

According to Web server logs, from 2010 to 2011 the PPFP e-toolkit received over 3,000 visits from 122 countries.

Practical Learning for a Geographically Dispersed Audience: The PPFP E-Learning Course

ACCESS-FP made use of another USAID-funded, electronic platform to offer an online course for global health professionals working on PPFP programs. The platform, called the Global Health eLearning Center (http://www.globalhealthlearning.org/), was sponsored by USAID in response to many requests from USAID field staff for updated technical information. With over 40 courses on global health topics, the eLearning Center emphasizes program learning—state-of-the-art technical information combined with program experiences and lessons learned presented in case studies (USAID, no date). USAID staff as well as other global health professionals use the eLearning Center. Experts from a variety of organizations author the courses.

Courses take approximately 2 to 3 hours to complete and include pre- and post-tests, as well as progress checkpoints (questions). Learners register for a free account to take courses and manage their progress. Learners who pass the final exam, submit an action plan of how they intend to use the course information in their work, and complete the course evaluation are allowed to print a certificate of completion. Approximately 2 months after completing the course, learners receive an automated e-mail (if they included a valid e-mail address with their registration) that includes their action plan and reminds them to continue steps to achieving the plan objectives. The eLearning Center has over 80,000 registered users, and more than 100,000 course completion certificates have been awarded.

Although the main delivery platform for the eLearning Center is the Web, selected courses have been distributed on flash drives and on CD-ROM. In addition, course materials may be printed out for offline use.

ACCESS-FP developed the PPFP course to orient global health professionals to service delivery, contraceptive methods, and programmatic considerations for family planning during the postpartum period. According to the eLearning Center administrative logs, over 1,300 learners visited the course at least once, and over 700 learners from more than 80 countries completed the certificate requirements for the course since its launch in 2008.

KNOWLEDGE TRANSLATION, SOCIAL NETWORKING, AND TEAM COLLABORATION CONCEPTS AS RELATED TO SCALING UP HIGH-IMPACT HEALTH PRACTICES

Of the knowledge management interventions that ACCESS-FP and MCHIP used to help programs scale-up high-impact health practices, the PPFP CoP provides the largest body of evidence for analysis. This evidence includes significant growth in CoP membership over time, anecdotal evidence of use of forum-communicated knowledge in programs, and a progression of activities arising from the CoP, such as formation of working groups to improve specific practices (e.g., LAM).

What makes a CoP effective? Recognizing that a CoP is characterized by sustained interaction and linkages among members, an examination of literature that focuses on interaction—knowledge translation, social networking, and team collaboration—may help provide answers to this question. These answers may, in turn, provide insights that can help programs become more effective in scaling up high-impact practices.

Defining Knowledge Translation

The term *knowledge translation* implies interaction between originators of research and implementers of healthcare programs and services.

A technical brief by the National Center for Dissemination and Disability Research (NCDDR, 2005) defines knowledge translation as active assessment and use of research knowledge to improve health practices and outcomes. Translating knowledge goes beyond just publishing research findings in journal articles; it extends to instituting process improvements and effecting behavior change in health systems (Davis et al., 2003).

The NCDDR brief summarizes several models for knowledge translation that were proposed in the literature. A framework proposed by Jacobson et al. (2003) guides design of a translation process through a series of questions about the user groups involved. Some of the questions and Jacobson's commentary on literature in the field provide insight into why the composition of the PPFP Community of Practice is well-suited to taking evidence to practice:

- *Centralized versus Noncentralized Decision-Making Power*: Starting with the technical consultation meeting in 2006, the PPFP CoP displayed broad sources of program decision-making power in its membership. Participants from more than 20 organizations and health programs analyzed research findings, formulated recommendations for programs, and took the recommendations back to their programs and another level of decision makers (e.g., project directors, country ministries of health). Jacobson notes that several studies linking the composition of user groups to knowledge translation suggest that groups in which the decision-making power is not limited to a small, central group are more likely to use research evidence. The large size of the CoP's online platform (the IBP Knowledge Gateway) broadened the base of decision makers who were exposed to the knowledge translation process.
- *Sophistication of the User Group's Knowledge of Research Methods and Terminology*: It has been shown that if research findings are to be used, they must match the user group's level of sophistication (Jacobson et al., 2003). The CoP's multitier communications helped filter PPFP program research findings. For example, through its comprehensive literature review, ACCESS-FP staff members provided the first refinement of research findings. Participants in the technical consultation meeting, who were sophisticated in assessing research, analyzed the filtered review results. Still more filtering and rephrasing took place when the mini-lectures were developed for the global online forum audiences who were more likely to be

program implementers than researchers. Finally, forum participants report passing along knowledge from the forums to health facilities and frontline health workers. Multiple levels of filtering and rephrasing can aid acceptance of the knowledge because the audience often received the knowledge from someone he or she knew professionally through a program or professional association.

- *Consensus on an Issue*: Drawing from the literature Jacobson notes that conflict or consensus on an issue and level of ambiguity affect knowledge translation by the user group. For this reason, the technical consultation meeting format was important for reaching consensus and reducing ambiguity on the PPFP program issues. The beginning plenary presentation of the literature review helped develop a common understanding; the later small group work to define recommendations aided consensus building, which would be important for later interaction among CoP members.

- *Dissemination Strategies*: According to Anderson et al. (1999) (cited in Jacobson et al., 2003), knowledge translation occurs through three key processes—awareness, communication, and interaction—whose forms need to be tailored to the preferences of the user groups for the amount and level of detail. The mix of knowledge management interventions used for PPFP attempted to touch on all of these key processes. Although a number of interventions were used—meetings, online CoP collaboration area and forums, and the electronic toolkit—surveys gather the most assessment data. For example, in one survey of global forum participants using the IBP Knowledge Gateway in 2008,* 42% of respondents reported forwarding forum messages to other people (IBP, 2009). The tendency to forward messages to colleagues supports the view that CoP interaction leads to a multitier dissemination and filtering of knowledge.

Another study suggests reasons why the CoP activities seem to be particularly effective in promoting change in PPFP programs. A randomized controlled trial (Dobbins et al., 2009) of 108 Canadian public health departments conducted in 2005 aimed to determine which are more effective as knowledge translation and exchange strategies—knowledge brokers, tailored and targeted messages, or an online research repository. The researchers viewed "knowledge brokers" as specialists who work

---

* Range of 200 to more than 760 participants; 11% to 14% response rate to survey.

one-on-one with decision makers to facilitate use of evidence in decision making. "Tailored" messages focus on a particular decision-making role, and "targeted" messages relate to a particular issue that requires a decision. An example of an online repository is a Web site with a searchable database of research findings.

The findings of the study again point to the importance of user group composition and culture in selecting knowledge translation and exchange strategies. For user groups whose culture placed a high value on evidence for decision making—and who were assumed to be more sophisticated in their knowledge of research methods—the labor-intensive role of the knowledge broker was less useful than tailored or targeted messages. For user groups with a low organizational research culture, the role of the knowledge broker was most helpful in encouraging use of evidence in decision making.

The study findings suggest a reason why the PPFP online forums may have gained increasing interest over time. Large global health donors such as USAID have a high organizational research culture, which leads them to give program funding to organizations that, similarly, have a high organizational research culture. Because many global online forum participants are based in these types of organizations (i.e., those that have a high research culture), the targeted and tailored messages of the forums would be appealing. Taking knowledge exchange one step further, some forum participants in turn work with healthcare facilities and government entities in low-resource settings to help them put evidence into practice. Whether these organizations have low or high research cultures would be a useful avenue for further research.

E-Learning Participants as Knowledge Brokers

According to the About page on the Global Health eLearning Center Web site, the courses blend technical content with program principles, best practices, and case studies in order to help learners improve global health programs (USAID n.d.). This approach is associated with the primary audience for the eLearning Center, which is USAID staff overseeing field programs, and the secondary audience, which is the staff of partner organizations implementing those programs, such as program officers.

In order to assess the usefulness of the courses to the target audiences, at the completion of courses learners are invited to complete a course evaluation. (Completing the evaluation is required in order to print the certificate.) One open text question on the evaluation form asks how the learner intends to use the course information in his or her work.

In course evaluations completed from 2008 to 2011, 258 respondents specified how they intended to use the course information. Given the goal of the eLearning Center to support program development and oversight, one would expect responses to the use question to predominantly mention improving programs. Instead, only 33 or 12.8% stated they intended to use the course information to improve programs, while 95 or 36.8% expressed intent to share knowledge from the course with colleagues, healthcare clients, the community, or participants of training or education. The remaining answers identified a variety of intended uses, such as improving his or her care of clients or furthering the learner's career.

Knowledge sharing, rather than program improvement, as the predominant intended use of the course presents a challenge in understanding the motivating factors for learners of the PPFP course. Perhaps most learners view themselves as knowledge brokers among their colleagues, clients, and other contacts, and so have a self-expectation to share knowledge that they think will benefit their contacts.

Learners may be sharing valued knowledge as a "gift" to other members of a community of practice in the hope that it will help someone and transform healthcare (Bell, 2010). (This type of "gift giving" is championed by Seth Godin [2008, 2010] who likens communities to "tribes" with strong ties and affiliation.) Viewing e-learning courses as a social networking tool, and so as a catalyst for knowledge exchange and translation, supports the approach of bundling e-learning with an online collaboration platform for communities of practice.

Because the literature shows that a key ingredient for successful knowledge sharing is trust (Holste and Fields, 2010), using multiple trusted sources supports widespread exchange of important knowledge. Worded another way, the use of a trusted, well-known knowledge source (USAID) via e-learning and knowledge brokers (learners) who are known and trusted by clients, policymakers, and program staff successfully creates a continuum of trust and knowledge sharing.

Social Networking Effects on PPFP Program Change

The entities involved in scaling up high-impact practices in PPFP span a wide variety of organizations and roles: international government and nongovernment organizations, public and private service delivery facilities and providers, education and training institutions, and others. Examining the linkages among these players in terms of social networking concepts provides insight into how the relationships may play a part in scaling up high-impact practices.

A growing body of literature examines public health systems as interorganizational networks. One such paper examined the relationship between governance, urbanization, public health domain, and health status in terms of social network structure (Wholey et al., 2009). The research found that collaboration among members of a network was more likely to occur when assessing or advocating for a particular public health problem than when seeking cofunding, unless a funder dictated it. This finding supports the technical consultation meeting format that (1) focuses on collaboration for problem solving among multiple organizations not tied to one funded program, and (2) "endorsement" for the event by a large funding organization.

Another study (Ramasamy et al., 2006), looking at relationship as a bridge to knowledge transfer, identifies three components for operationalizing a knowledge transfer relationship*:

- Trust
- Relationship commitment
- Communication

These three components play a part in the PPFP CoP interaction. Whenever possible, the facilitators of the global online forums identify contributors and their organizations in the daily digests. By identifying and crediting the sources of informal knowledge sharing, the facilitators help build relationships and trust.

Becoming a member of the CoP expresses a kind of relationship commitment to the domain that encourages knowledge exchange (Wenger, 2006; O'Brien and Richey, 2010). The CoP can be viewed as having just the right amount of structure for knowledge exchange: according to research (Ramasamy et al., 2006), too many rules for interaction (as might be defined in a contract) can stifle knowledge exchange, while too few rules can create ambiguity and inhibit exchange (e.g., a lack of rules may lead to questions about who is responsible for supplying knowledge and who

---

* Though the study specifically addressed Chinese business relationships, its findings agree with those for other settings.

should receive it). Regular communications through the IBP Knowledge Gateway to members, such as e-newsletters and announcements, maintain the sense of relationship.

Relationship plays a particularly important role in uptake of high-impact practices. A practice conveyed by a trusted colleague is more likely to be noticed and adopted than research findings that are published and disseminated widely. The role of opinion leaders in adoption of innovations is well known from Everett Rogers' *Diffusion of Innovations*, published in 1962. Yet other studies contend that any trusted contact, even if the contact does not fit the conventional image of a leader, can influence adoption. In one study (Williamson et al., 1989), researchers surveyed 625 physicians and 100 physician opinion leaders about their awareness and use of clinical advances. Even though physicians and opinion leaders said that publications were the most useful source of information on clinical advances, as many as one-fifth to one-half were not using or were not aware of the clinical advances identified in the study. This study found that practitioners actually relied more on colleagues and their own experience to validate information than they relied on published sources. This finding supports the importance of CoP interaction, particularly through the online forums, to drive adoption of high-impact practices.

Other research in social networking addresses the importance of how well an individual fills a "structural hole" between members of a network (Dekker and Hendriks, 2006; Nelson and Hsu, 2006). Individuals with ties that bridge gaps can serve as knowledge brokers and facilitate the flow of information and resources through the network (Nelson and Hsu, 2006). The concept of knowledge brokers as bridges that strengthen and extend networks supports the value of using a CoP platform (i.e., the IBP Knowledge Gateway) developed and endorsed by two major global health knowledge brokers: the World Health Organization (WHO) and USAID. WHO's and USAID's wide networks of contacts worldwide can generate more awareness for the work of a CoP than any other organization alone.

Team Collaboration and Macrocognition

Operationalizing the technical consultation meetings and CoP working groups requires use of effective team collaboration approaches. For the technical consultation meeting in 2006, in particular, it was important that participants work together as a team to analyze issues with PPFP programs and identify solutions for the way forward. A comparison of the 2006 technical consultation meeting format with stages described in macrocognitive process literature suggests components that can be replicated to ensure the success of future meetings.

Macrocognition describes the high-level mental processes that people use to solve complex problems in the real world, as opposed to problems presented in a controlled laboratory environment. Research on the macrocognitive processes of teams suggests they go through four stages of collaboration (Letsky et al., 2007). Table 5.2 compares the format of the PPFP technical consultation meetings with these stages.

In the Knowledge Construction stage, the meeting organizers arranged to debrief participants on the state of the literature regarding PPFP programming. Because the meeting was intended to be a practical, team-driven activity (rather than a general knowledge update frequently provided by conferences), it was important for organizers to identify the knowledge threshold they wanted individual participants to achieve by the end of this stage. This threshold was then the basis for the Collaborative Team Problem-Solving stage, during which participants broke into small groups and began functioning as teams to assess portions of the literature review findings. Organizers needed to be prepared with a plan for how to frame the problem so that it could be addressed with a team approach. By the end of the small group analysis, the participants needed to come together to form one group during the Team Consensus stage for the purpose of identifying the specific action plan that would result from the meeting. Finally, the organizers needed to plan for follow-up on the action plans during the Evaluation and Revision stage, with checkpoints at annual meetings.

TABLE 5.2

The Technical Consultation Meeting Can be Compared to a Macrocognitive Process Format

Technical Consultation Meeting Feature	Macrocognition/Team Stage
Technical update presentations	Knowledge construction (individual knowledge building)
Small-group analysis of literature review	Collaborative team problem solving
Small-group recommendations based on literature review findings	Team consensus
Plans for follow-up on recommendations and annual meetings	Evaluation and revision

LESSONS LEARNED

Large global health programs with defined knowledge management strategies face some of the same challenges that individuals in public health systems face; that is, there is an abundance of published information to filter, and the best knowledge is often in someone's head. A coordinating entity, such as the ACCESS-FP program staff, can help with filtering information through literature reviews and electronic toolkits. However, tapping into the "know-how" and lessons learned of health professionals working in low-resource settings is a greater challenge. In these settings, time—to document experiences, evidence of best practices, and success stories—is usually not allocated as part of health professionals' jobs. These health professionals may not have access to or awareness of as many publishing opportunities as colleagues who work in regions with reliable Internet connectivity and access to scientific publications. Global health programs need to explore innovative ways of collecting and disseminating know-how from the field—other than the traditional journal articles and conference presentations. Use of consumer products such as mobile phones (to record mini-podcasts) or flip cameras (to capture video for success stories) may offer solutions.

For global health programs with worldwide partners, language barriers may make it difficult to encourage interaction. A perennial problem for discussion forums, for example, is whether to hold separate forums by language, rather than by topic. Use of more sophisticated translation tools may help solve this dilemma in the future.

In discussing the case study, relatively little was stated about using information from the PPFP electronic toolkit to improve PPFP programs. For all global health information repositories, identifying how the information was used in programs is a challenge. Typically, Web-based surveys are used to collect this type of information, but the results give only a partial view of use. Cross-promotion and cross-evaluation of information services can help obtain a broader sense of how the information was used.

Finally, a discussion of global health programs would not be complete without asking, "What about sustainability?" When developing a knowledge management strategy for a global health program, it is important to ask, "How will users access that valuable knowledge after a funded program is ended?" Plans for hosting Web sites, storing knowledge in repositories, and archiving discussions need to be made early in the program to ensure a life cycle that matches the global health domain, not just the life of the program.

FUTURE RESEARCH NEEDED

Social networking analysis of global health programs could go a long way toward answering the question: "How did this program improve health practices?" During the course of a large global health program, many organizations and individuals may be exposed to the program's knowledge outputs through regional meetings, online discussions and collaboration, Web sites, electronic learning courses, and other means. Research to track the knowledge exchange and translation "trail" from organization to organization and person to person would help to develop evidence of which knowledge management approaches work best to scale up awareness and use of high-impact health practices.

ACKNOWLEDGMENTS

I would like to thank Jean Sack for research assistance and Christine Merriman for editorial assistance. I would also like to thank the following colleagues for sharing information about their work on the MCHIP, ACCESS, and ACCESS-FP programs: Samaila Yusuf, Catharine McKaig, Elizabeth Sasser, Barbara Deller, Jeffrey Smith, Nancy Caiola, Koki Agarwal, Steve Hodgins, Leo Ryan, Christine Merriman, Jaime Mungia, and Charlene Reynolds. Finally, I would like to thank the Knowledge4Health program at the Johns Hopkins Bloomberg School of Public Health, Center for Communication Programs, for permission to show the design of the PPFP e-toolkit.

REFERENCES

ACCESS-FP. 2010. *The Evolution of Postpartum Family Planning—Meeting Report.* May 13, 2010, Washington, DC.

ACCESS-FP. 2007. *Postpartum Family Planning Community of Practice.* Baltimore, MD, and Washington, DC: ACCESS-FP and USAID. Accessed July 18, 2011, at: http://www.k4health.org/system/files/Postpartum%20fp%20community%20of%20practice.pdf

ACCESS-FP. 2006. *Postpartum Family Planning Technical Consultation—Report Brief.* Washington, DC. Accessed July 18, 2011, at: http://www.accesstohealth.org/toolres/pdfs/ACCESSFP_PPFPrptbrief_nov06tc.pdf

Action for Global Health. 2011. *Aid Effectiveness for Health: Towards the 4th High-Level Forum, Busan 2011: Making Health Aid Work Better.* Accessed September 5, 2011, at: http://www.actionforglobalhealth.eu/fileadmin/AfGH_Intranet/AFGH/Publications/2011_Policy_Report_-_Aid_Effectiveness_for_Health/AFGH__FINAL___WEB_.pdf

Advisory Committee on Voluntary Foreign Aid (ACVFA). 2007. *ACVFA's Analysis and Recommendations of Trends in USAID Implementation Mechanisms.* Washington, DC: USAID. Accessed September 5, 2011, at: http://pdf.usaid.gov/pdf_docs/PDACJ790.pdf

Anderson, M. et al. 1999. The use of research in local health service agencies. *Social Science and Medicine* 49(8): 1007–1019.

Bell, J. 2010. Social media and family nursing: Where is my tribe? *Journal of Family Nursing* 16(3): 251–255.

Campbell, O.M., and Graham, W.J. 2006. Strategies for reducing maternal mortality: Getting on with what works. *Lancet* 368(9543): 1284–1299.

Cleland, J., et al. 2006. Family planning: The unfinished agenda. *Lancet* 368(9549): 1810–1827.

Davis, D. et al. 2003. The case for knowledge translation: Shortening the journey from evidence to effect. *British Medical Journal* 327(7405): 33–35.

Dekker, D.J., and Hendriks, P.H.J. 2006. Social network analysis. In *Encyclopedia of Knowledge Management*, D. Schwartz (ed.), pp. 818–825. Hershey, PA: Idea Group.

Dobbins, M. et al. 2009. A randomized controlled trial evaluating the impact of knowledge translation and exchange strategies. *Implementation Science* 4: 61.

Esser, D.E. 2009. More money, less cure: Why global health assistance needs restructuring. *Ethics and International Affairs* 23(3): 225–234. Accessed September 5, 2011, at: http://www.carnegiecouncil.org/resources/journal/23_3/essays/001

Godin, S. 2008. *Tribes: We Need You to Lead Us.* New York: Penguin.

Godin, S. 2010. *Linchpin: Are You Indispensable?* New York: Penguin.

Holste, J.S., and Fields, D. Trust and tacit knowledge sharing and use. *Journal of Knowledge Management* 14(1): 128–140.

Implementing Best Practices (IBP) Initiative in Reproductive Health. 2009. *IBP Knowledge Gateway: Creating Virtual Knowledge Networks to Share and Exchange Information.* Poster. Accessed September 11, 2011, at: http://www.who.int/reproductivehealth/publications/Poster_IBP_2009.pdf

Jacobson, N., Butterill, D., and Goering, P. 2003. Development of a framework for knowledge translation: Understanding user context. *Journal of Health Services Research and Policy* 8(2): 94–99.

Letsky, M. et al. 2007. Macrocognition in complex team problem solving. In *Proceedings of the 12th International Command and Control Research and Technology Symposium*, June 19–21: 2007, Bellevue, WA.

MCHIP. No date. *MCHIP Vision and Strategy.* Accessed September 8, 2011 at: http://www.mchip.net.

National Center for the Dissemination of Disability Research (NCDDR). 2005. What Is Knowledge Translation? *Focus* Technical Brief, Number 10. Accessed September 9, 2011, at: http://www.ncddr.org/kt/products/focus/focus10/Focus10.pdf

Nelson, R.E., and Hsu, H.Y.S. 2006. A social network perspective on knowledge management. In *Encyclopedia of Knowledge Management*, D. Schwartz, ed., pp. 826–832. Hershey, PA: Idea Group.

O'Brien, M., and Richey, C. 2010. Knowledge networking for family planning: The potential for virtual communities of practice to move forward the global reproductive health agenda. *Knowledge Management and E-Learning: An International Journal* 2(2): 109–121.

Ramasamy, B., Goh, K.W., and Yeung, M.C.H. 2006. Is Guanxi (relationship) a bridge to knowledge transfer? *Journal of Business Research* 59(1): 130–139.

Ringheim, K. 2011. *Integrating Family Planning and Maternal and Child Health Services: History Reveals a Winning Combination*. Washington, DC: Population Reference Bureau. Accessed September 8, 2011, at: http://www.prb.org/Articles/2011/family-planning-maternal-child-health.aspx

Rogers, E.M. 1962. *Diffusion of Innovations*. New York: Free Press of Glencoe.

Ross, J.A., and Winfrey, W.L. 2001. Contraceptive use, intention to use, and unmet need during the extended postpartum period, *International Family Planning Perspectives* 27(1): 20–27.

U.S. Agency for International Development (USAID). 2011. *High Impact Practices in Family Planning*. Accessed September 6, 2011, at: http://www.usaid.gov/our_work/global_health/pop/publications/docs/high_impact_practices.pdf

USAID. 2006. *User's Guide to USAID/Washington Health Programs*. Washington, DC: USAID. Accessed September 8, 2011, at: http://pdf.usaid.gov/pdf_docs/pnadg713.pdf

USAID. No date. *Global Health eLearning Center Fact Sheet*. Accessed September 14, 2011, at: http://www.globalhealthlearning.org/learnmore.cfm

Wenger, E. 2006. *Communities of Practice: A Brief Introduction*. Accessed September 11, 2011, at: http://www.ewenger.com/theory/index.htm

Wholey, D.R., Gregg, W., and Moscovice, I. 2009. Public health systems: A social networks perspective. *Health Services Research* Part II, 44(5): 1842–1862.

Williamson, J.W. et al. 1989. Health science information management and continuing education of physicians: A survey of U.S. primary care practitioners and their opinion leaders. *Annals of Internal Medicine* 110(2): 151–160.

6

Emperor: A Method for Collaborative Experience Management

Ulrike Becker-Kornstaedt and Forrest Shull

INTRODUCTION

Software developers have an array of methods, tools, and techniques—generically, "practices"—to choose from in tackling the complexities of engineering good software. Practice choices can range from large-scope issues that will impact almost all the other decisions on a project (such as whether to use an iterative life cycle model to organize the work) to fine-grained decisions such as whether to use a specific testing approach. In addition to the suite of well-tested practices already in existence, new practices are being created all the time in an attempt to keep up with the ever-growing needs of contemporary software teams. Software developers are bombarded by these practices all the time in books, magazines, podcasts, and even in the scientific literature, yet remarkably little information is easily available that describes the usefulness of a particular practice in any given situation.

The difficulty in choosing the "right" practice for a given situation is that the success—or failure—of a practice often is not solely rooted in the practice itself, but in applying the right practice for the right context. A practice that was highly successful in one environment may be a complete disaster in a different one. Adopting a new practice, on the other hand, can mean a considerable investment in terms of cost and effort (e.g., training, overcoming the learning curve).

To help users of a practice select those practices that are suitable candidates for their current situation, additional information on the practice is necessary, such as past successes or failures as well as the contexts in which they occurred. A good indicator of whether a practice will be useful in a given context is past experience with it in similar contexts. A wide range of experience is needed to get a rich picture of a practice to allow others to make an informed decision about the potential adoption of a practice. However, this experience typically sits in the heads of different types of individuals or is scattered across different data sources in different locations and needs to be mined using adequate techniques. When experience reports about practices already exist or can be elicited, they are not always reliable enough to base such a critical decision upon. Experience reports may contain incorrect, incomplete, or misleading information. For instance, an experience report may claim to have used a new, hot-button practice but in fact left out key components; may provide a biased view by reporting only benefits while leaving out costs; or may leave out important aspects such as the fact that the goal was saving costs, even at the expense of quality. Results might not have been put into the bigger picture, when the success of applying the practice may be due to other activities going on at the same time. In addition, experience reports are not always understandable for people who were not involved in applying the practice.

To provide valuable and validated experience to users in domains like this, the EMPEROR approach (Experience Management Portal using Empirical Results as Organizational Resources) (Shaw et al., 2007; Shull et al., 2006) for distributed evidence handling was developed. EMPEROR envisions a system that collects experience reports on practice applications shared by practitioners. Experience is handled, summarized, and interpreted by experts worldwide, who are working collaboratively and constructively, and are tapped according to their experience. To find a suitable practice, a potential user would enter information describing his or her current context, such as company or team characteristics and application domain and desired outcome of applying a practice, such as increased quality or reduced cost. As a search result the user would get a list of practices that yielded similar results in comparable contexts. Because "perfect" content will not always be available, the system can be iteratively expanded and improved, incorporating content as it becomes available. To take the iterative nature of the system into account, content will be annotated according to its maturity.

A key component is relying on experts to help put experience reports into context and ensure quality, so that confidence can be had in decision making. However, because the necessary experts for such an endeavor are dispersed and their time is limited, EMPEROR provides clear processes, terminology, and templates for consistently capturing and maturing evidence, which can efficiently support a distributed community. Using such a coordinated approach minimizes experts' effort for handling evidence, which also makes it feasible to get the support of outside experts if the required expertise is not available within the organization.

This chapter discusses the EMPEROR approach and its focus on allowing dispersed domain experts to collaborate on knowledge creation. The second section illustrates the background of the organization and the motivation that led to the design of the EMPEROR method. Based upon a literature review on social networking and collaborative content handling, the third section discusses the principles used to design EMPEROR and describes how these principles were applied in the design of the method and associated processes. Specifically, to illustrate the key collaborative principles, we will focus on the key process of "evidence handling," which dispositions suggestions from multiple sources regarding evidence that should be captured for decision making, and sketch two more processes that deal with aggregation and vetting of practice information.

An application of EMPEROR in a case study at the Defense Acquisition University (DAU) and the lessons learned from the case study are described in the fourth and fifth sections. The last section summarizes the chapter and points out areas of future research.

BACKGROUND OF THE ORGANIZATION

EMPEROR was developed at the Fraunhofer Center for Experimental Software Engineering (CESE), a not-for-profit applied research and technology transfer organization founded in 1997. CESE aims to apply empirical research methods to real-world problems in software and system development and to transfer practical and novel solutions to industry. Across a variety of contexts, CESE applies competencies in software measurement to provide insight of what is truly happening during software development, and competencies in knowledge management to encapsulate that understanding so that other projects may benefit.

CESE Knowledge Management services are based on the *experience factory* (EF) approach developed by Basili et al. (1994), one of the foundational approaches for making use of experiences to learn from past failures and successes and improve software engineering practice.

An organization that has an EF in place not only stores concrete project artifacts, such as code modules, in a project repository, but in addition stores experience gained in software projects in an experience base (EB) in the form of models—abstractions of the activities, artifacts, or objects used in the project. An activity model for code review, for instance, would describe steps taken for that activity, and could include templates used for code reviews or describe preconditions for the code reviews. Each model in the EB is associated with a characterization of the project contexts where the model was used. For reuse, the models with the context that best matches the new project context are selected from the EB, adapted, and used in the new project. After its use, data on model usage is analyzed, and the model in the EB is updated to reflect the experience gained with its new application. The EF concept was successfully implemented in various industry contexts, such as NASA (Basili et al., 1992) and Daimler (Houdek et al., 1998).

COLLABORATION AND SOCIAL NETWORKING PRINCIPLES APPLIED

EMPEROR is aimed at building evidence-based decision support for practice selection. EMPEROR is based upon the EF principles in that it builds content based on experience and helps the next project find the right content based on the current situation. Implementing a new practice into a given context can be a major change for an organization and can have considerable implications in terms of cost and effort. The cost of process change and sometimes long latencies of implementing a new practice make it necessary to select the right practice from the beginning. Thus, a major goal of EMPEROR is the *trustability* of its content—the information on which users will base the decision of whether to adopt a new practice. Two key aspects of trustability are *completeness*, ensuring that known effects of a practice are described, and *correctness*, meaning that the effects of the practice are correctly recorded and important information is not left out. To appropriately handle and interpret experience and information that may be incorrect, inconsistent, or incomplete and eventually transform it into evidence records requires experts with the practice. Since their time is often limited, a second goal of EMPEROR is to coordinate experts in a way that they can efficiently and collaboratively handle such content. Third, we want people to use EMPEROR. This relates to the usefulness of the information provided by the system, and to its usability.

In designing processes to enable experts to collaborate effectively, we reviewed the literature and similar collaborative networks aiming at generating or compiling content to select guiding principles for our own work. EMPEROR is built on the experience factory concepts as it builds content based on experience and helps find the right content based upon the current context. However, whereas an experience factory is typically established within an organization, EMPEROR works across organizations. The EF focuses on concepts for selecting and storing experience; EMPEROR aggregates existing experience into new content.

Selected Principles for This Context

The goal of EMPEROR is to build high-quality evidence through the efficient coordination of experts handling experience with practices. Collaborative principles to build, collect, or compile knowledge or information are used in many different contexts, in different organizations or communities, and with different goals and purposes. To support collaborative work, software was developed. For EMPEROR, we review concepts and principles from related approaches to build upon.

In reviewing similar initiatives that compile and aggregate specialized domain knowledge to support decision makers, the Cochrane Collaboration is among the best matches to the intents of EMPEROR. The Cochrane Collaboration aims at supporting decision making in healthcare by collecting, reviewing, and summarizing evidence on the healthcare domain (The Cochrane Collaboration, 2011). Content provided is largely compiled and reviewed by experts, which generally results in higher maturity of content. The goal of the Cochrane Collaboration—achieving high-quality content—is similar to the goals of EMPEROR, making these principles good models for EMPEROR.

Companies working in the knowledge management space, such as Mongoose Technology and Dave Pollard (independent consultant), present some basic principles for tools to support social networking and group collaboration (Pollard, 2005). EMPEROR is not focused on networking per se, but uses networking concepts to build its processes upon. Thus, basic principles on how people interact with each other and are coordinated are helpful as they help frame collaboration in EMPEROR.

The online encyclopedia Wikipedia is a widely known example of how a large amount of content is contributed and maintained collaboratively by a large community of anonymous volunteers. In Wikipedia anyone can add or modify or delete any content. A core principle of Wikipedia is its verifiability; however, major criticism of Wikipedia addresses the lack of reliability or trustability of its content. Wikipedia content can range anywhere from highly trusted and reviewed information to incorrect and unsolicited information (see, for instance, Seigenthaler, 2011). The content submission principles and the wide range of contributors of Wikipedia make great models for EMPEROR, but their content handling principles lack the rigor needed to ensure high-quality content and efficient handling of content.

Retail Web sites such as Amazon.com often show customers feedback on items purchased, and sites such as yelp.com provide review systems as their business model. Some of these sites even encourage other users to rate the helpfulness of the review or allow people to dispute reviews. Disagreeing reviews typically lead to more detailed and interesting information, thus increasing the content base. A high helpfulness ranking of reviews and a high number of reviews increase an Amazon reviewer's rank and credibility. However, in spite of a feedback-on-feedback mechanism, reviews may not always be objective (Pinch et al., 2011). For instance, a negative made-up review from a competitor or positive reviews from a friend asked for a favor can give an inaccurate picture of the item being reviewed, and because users of the site can be hurt by inaccurate reviews, a skeptical eye may sometimes be in order. Many sites use a ranking system to offer an overall recommendation across the reviews, aggregated from the individual rankings of the item.

Based upon analyzing these examples, and considering the goals of EMPEROR—trustability, efficient expert collaboration, and usability—we used the following as guiding principles for EMPEROR. The first set of principles relate to the people providing and handling content.

> Principle 1: *Enable wide participation from a wide range of content providers*: To obtain a reliable, comprehensive, and rich picture of a practice and possible effects of its application, not just a large number of experience reports are needed. Contributions from people with different skill levels and which cover diverse contexts are required and have to be encouraged and facilitated.

Principle 2: *A wide range of experts to handle content*: Handling a wide range of contributions requires experts in these fields to adequately assess and process these contributions.

Principle 3: *Coordination of experts*: Good management and coordination of experts are needed to take into account the limited availability of experts and to make optimal use of their time.

The next set of principles relates to the trustability of the content presented by the system:

Principle 4: *Provide high-quality content*: The quality of content is key to the system; thus quality needs to be achieved by incorporating rigorous quality assurance methods into the method.

Principle 5: *Build the reputation of information handlers*: Information handlers need to have or build a good reputation and credibility. This increases the quality of the content presented and the confidence that the users have in the system and the information provided by it.

Principle 6: *Assess the relevance of published content*: Content needs to be relevant and interesting in order to provide useful and helpful information and keep the system in use. Irrelevant information can result in users making wrong decisions.

Principle 7: *Keep content up to date*: Outdated information may lead users to select inappropriate practices for their context. Thus, content needs to be kept up to date through periodic reevaluation of existing content and through the identification and incorporation of new content as it becomes available.

Principle 8: *Minimize bias*: Experience reports reflect the view of the content provider or the context where the experience was gained. To establish independence and trustability of content, a variety of approaches such as internal vetting of the content, ensuring broad participation, and avoiding conflicts of interest need to be taken.

The last subset of principles describe the ways users will find and use the content that is output of the method:

Principle 9: *Provide indicators of content trustability*: Evidence that does not have the desired quality may still need to be incorporated (e.g., if it is the only evidence available). To make users aware of potentially less trustable content it needs to be marked as such.

Principle 10: *Cater to a wide range of users*: If the resulting system is targeted toward users from different backgrounds or skill sets, this needs to be taken into account. This can be done, for instance, by providing multiple ways of organizing the content and multiple levels of granularity of results.

Principle 11: *Provide users with an experience they can relate to*: People are more willing to accept advice if they can tie it to their own experience. Often, people learn better from listening to other people's stories than from a "how-to" lecture from an expert. This may also have an impact on the way content is presented and the media format used.

Principle 12: *Provide aggregated information as well as details*: With content built up from small pieces there is an inherent risk of users losing the overall picture. Aggregating information helps users maintain an overview in order to avoid being overwhelmed by too much information.

Incorporating Principles into Processes

EMPEROR defines a context-specific template for capturing evidence, each record of which must be linked to a published practice and have passed a rigorous evaluation and well-defined quality assurance process. In EMPEROR, a *practice* is a repeatable activity, defined in such a way that someone other than the definer can implement it with demonstrable consistency. An EMPEROR *evidence* record contains detailed information about one environment in which the practice was used (so future users can decide whether the evidence is likely to apply to them) and a description of the result, such as impact of the practice on cost, effort, or quality. To gain access to the expertise needed for building up content, EMPEROR distributes the handling of practices and evidence prior to publishing among several subject matter experts (SMEs), coordinated by a content manager (CM). Typically, an SME is a known expert in the field (fulfilling *Principle 1*) who should be trained on the EMPEROR method and whose reputation can be conveyed along with the content that expert edited (*Principle 5*). As domain experts, SMEs are familiar with the latest trends in their field of practice and can provide additional information that can help users when learning more or implementing the practice (*Principle 11*).

For EMPEROR four processes were defined: internal content handling, writing practice summaries, practice vetting, and content maintenance.

The process for evidence handling is depicted in Figure 6.1. A content provider submits an experience report to the system. In most cases, the content provider was involved in applying the practice, but we also devised methods for experience capture and collecting content, such as document analysis and interview techniques tailored specifically to the EMPEROR definitions of practice and evidence. These can be applied to facilitate wider participation (*Principle 1*). Evidence submitted to the experience base (EB) is prescreened by the CM, who also performs a brief sanity check. The goal of this prescreen is twofold: first, it ensures that evidence submitted fulfills minimal quality criteria so the SME can stay focused and productive at all times. Second, prescreening allows the CM to assign the piece of evidence to the SME who is best suited to handle it (*Principle 2*). All SME activities related to an entity are coordinated by the CM, which avoids duplication of work and thus optimally uses experts' time (*Principle 3*).

Once an evidence record is assigned to an SME, it is reviewed according to a set of predefined quality criteria (*Principle 4*) to assess accuracy, understandability, completeness, and relevance (*Principle 6*). The SME may make some changes or additions so the evidence meets these quality criteria. One quality aspect, for instance, requires any identifying information to be removed when a content provider chooses to remain anonymous. In order to provide valuable information to users while protecting information providers and to encourage them to submit content in the future, sensitive information that can harm the information provider or the organization where the experience was gained (Becker-Kornstaedt, 2001) needs to be treated. In some cases the SME may even have to consult additional information sources or contact the person who provided the evidence for further information. Based on the review of the evidence, the SME makes a decision regarding what to do with the evidence:

- If a different SME would be better suited to handle the evidence, the evidence is reassigned to that SME. This avoids effort being spent unnecessarily (*Principle 3*) while ensuring that content is handled by the person optimally suited for it to ensure quality (*Principle 4*).
- An SME might find that the submitted evidence is not useful and will not ever be useful (e.g., the evidence is outdated or, even if enhanced, cannot be useful evidence for users). Thus, the evidence is rejected and will eventually be removed from the system by the CM.
- In situations where it is not yet clear to the SME how to process the evidence, he or she can defer the evidence to a lead. A lead is

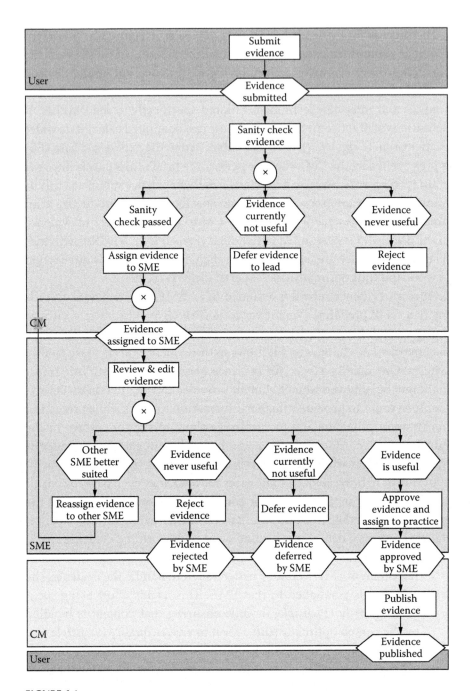

FIGURE 6.1

Evidence handling in EMPEROR (Experience Management Portal using Empirical Results as Organizational Resources).

potential evidence that currently cannot be processed (e.g., because an SME with specific expertise is missing, or because the content is not a focus area but may become one). As opposed to the previous alternative, information is removed from the SME's work queue but remains within the system for revisiting at a later time.

- If an evidence record has sufficient quality to go public, the SME can publish it, and the evidence is publicly available to the users of the EB. Assessment schemes for trustability were developed for evidence information and practice information (*Principle 10*). The ranking scheme for evidence takes into account factors such as how evidence was gained and how reliable that information is.

Handling practice suggestions follows a similar process by their assigned subject matter experts (SMEs).

Once a practice accumulated enough quality evidence, it is assigned to its SME to write a summary. The SME rechecks all the fields of the practice, ensures that enough and adequate supporting information on the practice description is available and still up-to-date. All of the available evidence, together with the SME's professional experience, is then aggregated into an easy-to-read summary of key information for the practice (Principle 13). The summary lists information such as the overall benefit of using the practice, and the net impact on project cost, quality, and schedule. The summary can include additional information or risks (e.g., warnings about side effects). This summary will help potential users make a quick assessment on whether this practice is suitable for their current situation without having to browse through all the individual evidence records. In addition, the SME adds information, such as guidelines for monitoring the practice, or references to standards with which this practice complies.

The *vetting* process for practice information coordinates a team of independent experts performing a final objective check of the content to ensure completeness, nonbias, or usefulness of information. Like the SMEs and CMs, vetters examine the quality of each piece of practice and evidence information with respect to whether it is clear and makes sense. Unlike these other roles, however, vetters also look at the accumulated body of content on a topic to ensure that all viewpoints and all known applications are taken into account. Having three levels of quality ensures high quality of all content published (*Principle 4*). Having content reviewed by an independent vetting team, representing different viewpoints and application areas, ensures completeness and relevance (*Principle 6*) and helps avoid bias (*Principle 8*).

These three processes reflect the three levels of overall maturity and trustability of practice information. Each practice is clearly labeled as to its progress through these processes (Shaw et al., 2007). A practice that passed the basic handling process is marked as "bronze," a process that has a summary written is marked as "silver," and a practice that passed the vetting process is marked as "gold," visualizing maturity of content to users (*Principle 10*). All EMPEROR processes are defined so that the CM provides overall coordination and SMEs provide specific domain expertise (*Principle 3*). The CM filters content before it is assigned to an SME. The CM has the final decision on rejected and deferred content and triggers summary, vetting, and maintenance processes. This ensures that an SME hardly ever deals with evidence outside his area of expertise or with evidence that cannot be appropriately processed, and that information cannot be removed from the underlying experience base without CM approval.

In addition to these processes, which generate new information, a basic *content maintenance* process was defined for EMPEROR to ensure that content already published remains up to date, new information that is integrated into the system is consistent with existing information, and that content conforms to state of practice. Together with some quality aspects checked as part of initial content handling, the maintenance process ensures that information in the system is up-to-date (*Principle 7*). Throughout the life cycle of an entity, the CM is in charge of the history of each entity, allowing for continuity (*Principle 9*).

EMPEROR is not tied to a specific tool, but due to the nature of these processes, a tool that coordinates distributed users is recommended to support these processes.

DESCRIPTION OF CASE STUDY

EMPEROR was applied as part of the U.S. Department of Defense's (DoD) knowledge management efforts. This section details the key aspects of how the EMPEROR concepts were implemented in that environment. Section 804 of the National Defense Authorization Act of 2003 (Data and Analysis Center for Software, 2011) directed the Office of the Secretary of Defense (OSD) to establish a clearinghouse for best practices in software development and acquisition. As a result, the DoD best practices clearinghouse (BPCh) (Dangle et al., 2005) was established at the Defense Acquisition University (DAU).

DAU is an institution that offers and manages training, learning, and development opportunities for U.S. Department of Defense members and their partners in the defense industry through courses and knowledge-sharing tools. Thus, potential contributors, users, and even SMEs are located in different parts of the world, requiring an asynchronous and distributed process.

Tool Support for Collaborative Work

To support the process and its distributed users, DAU tool support was crucial. The BPCh tool was implemented using a Microsoft SharePoint (Sharepoint) installation with a Web-based interface, facilitating the coordination of SMEs and CMs who are geographically distributed and will work asynchronously. Each SME can access the system from any location in the World Wide Web using a personalized account to process and evaluate the submitted content and route it through the system to make it publicly available for the end user community. BPCh SMEs are experts in the respective field of the practice, who in addition are trained with the processes and the tool. A sample practice description in BPCh is shown in Figure 6.2. In this interface, a BPCh practice consists of a description and a summary. The right-hand side of the practice view lists resources and

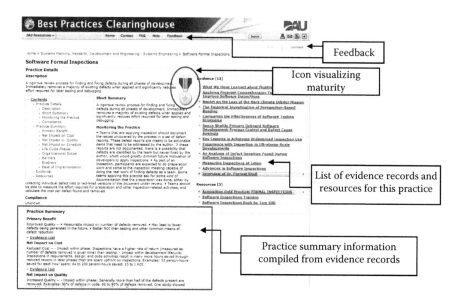

FIGURE 6.2
Screenshot of practice in best practices clearinghouse (BPCh) tool. (https://bpch.dau.mil, last accessed September 14, 2011, with Safari browser.)

evidence available for a practice. These are linked to the corresponding records, allowing users to view their detail as needed. Resources provide pointers to any additional supporting information that may help a user perform a practice (Principle 11).

A bronze, silver, or gold medal icon, respectively, visualizes practice maturity (Principle 10). For the silver summary, information such as primary benefit, net impact on cost, quality, and schedule is compiled across all evidence records for that practice, as seen in the lower left. Feedback buttons or tabs allow users to leave comments on the practice or the overall system.

As of 2010, evidence for about 60 practices was evaluated, routed through the system, and made publicly available. Additionally, more than 200 evidence candidates were submitted to the tool and processed in the back-end of the tool. Practices can be browsed by career field, benefit, or keyword.

Quality

The generic list of quality criteria developed for the EMPEROR method was adapted to the specific situation at BPCh and refined into a checklist, to make quality-checking more operational. This checklist ensures a consistent understanding of quality criteria across the different experts and ensures the quality of the resulting output. We view the checklist as a living document that will be refined and adapted based on feedback over time as more experience on content handling becomes available. In addition to internal quality checking mechanisms on the SME side, external users can provide feedback to the system or content of the system using feedback buttons in the user interface.

Trustability

Trustability of an individual evidence record is rated using a 20-point scale (Shull et al., 2006). The scale takes into account the completeness of the available data and the trustworthiness of the evidence. Points are accumulated based on independent scores for how the evidence was reported, the objectivity of the person who reported the evidence, how the practice was applied, and how the results were measured. If the evidence, for instance, is based on a vendor reporting on his Web site that his tool was used in a project, it would be rated with 5 points ($1 + 0 + 4 + 0$). A publication in a book or highest-quality archived journal, written by a member of the team, reporting on the application of the practice in a series of production projects with a detailed measurement data comparing the results to existing data would score 20 points (5 + 3 + 7 + 5) (see Shull et al., 2005, for details). The trustability score is shown clearly on the interface. In addition to showing evidence trustability, the score is used to trigger the practice summary process: as soon as the accumulated evidence for a practice reaches 50 points, the summary can be written.

Traceability

To maintain a history of each piece of content, in each step of the routing process the SME is asked to provide a comment that will be kept in the internal part of the database. Comments, for instance, why an SME judged the evidence to be useless, help the CM in tracing the path evidence took, and allowing to document and retrieve a detailed processing and decision history for each piece of content handled. To keep content providers involved and interested, they are informed of the state of their contributions and any status change (e.g., evidence that goes public, practice that gets promoted from silver to gold) is communicated to them via e-mail.

Seeding the System

To give users a feeling of content structure, and to encourage potential content providers to submit their experience, the system needed to have useful content from its launch. Thus, BPCh was seeded with content that was seen as relevant to the target domain. To do so, experience reports related to the defense domain or from contexts similar to the target domain were processed to jump-start the system with practice information.

Blitz Interviews

To reach out to a wider circle of content contributors, we conducted blitz interviews—highly structured interviews of about 5 to 10 minutes duration, following the BPCh evidence format, at one of the flagship conferences of the defense system engineering community. Using a proactive interviewing approach allowed getting experience reports from people who might not have taken the time to use the Web interface provided by the tool. The interviews turned out to be an easy and efficient way to reach a wide range of experts to tap into their experience. In addition, video clips taken of the interviews could be integrated into BPCh. These clips turned out not only to be a great medium to supplement existing (written) evidence, but they also made experience reports livelier than their transcribed counterparts (*Principle 12*).

Vetting

BPCh targets users from military/government and industry. To take into account these different viewpoints of these users, BPCh vetting teams consist of representatives from government, industry, plus a member from academia to ensure that new trends and latest findings from research are considered.

Training

While our SMEs were experts in their respective fields, they had to learn how to handle content with the tool. Thus, we developed training modules describing content handling processes. Following the distributed location of the SMEs, the training was administered in the form of a Webinar with participants in different physical locations. In addition, videotaped training sessions were made available to SMEs via the BPCh Web site. Training was supplemented by coaching the new SMEs while handling their first couple of items. This proved to be an efficient concept in bringing new SMEs up to speed while also providing feedback on the underlying processes and the supporting system.

LESSONS LEARNED

Implementing the EMPEROR in the DAU BPCh showed the feasibility of our approach for collaborative decision handling and helped us gain experience on issues to consider when implementing such a system.

Deal with imperfect evidence—even though anticipated, it is often necessary to deal with less-than-perfect content. In particular for new practices, it is difficult to get high-quality evidence. In such situations unreliable content is better than no content, but it needs to be explicitly marked as such. The ranking scheme helps to indicate the trustability of the content for potential users.

Jump-starting the EB with content borrowed from other contexts before going public is important to ensure that users will use the system. However, the context of the seeded content has to match the target context as closely as possible to provide valuable information from which users can draw appropriate conclusions for their own situations. Also, the borrowed content may have to be weeded out once enough proprietary content is available, in order to match the target environment as closely as possible.

Thoroughly defining or tailoring the data structures and the fields for evidence and practice records to the specific application is crucial. This includes defining how different fields will be used and which additional fields will be needed in the different steps of the overall life cycle of the practices and evidence records. A clear definition of the data structures makes it easier for a potential content provider to submit the information and provide the quality and content needed. It also aids the SMEs in handling or processing information and eventually may help to partly automate steps in the handling processes.

Training SMEs is important, not only because it enables them to handle content, but we also realized that our SMEs got excited about the project as they get to see the bigger picture and the importance of their role in it. Our mentoring approach was very efficient in bringing SMEs up to speed. Mentoring and even shadowing SMEs while handling content records worked particularly well as the actual routing task requires relatively little time, but waiting for an answer to a question that might come up during the execution of the process might have put off the SMEs. In addition, shadowing the SMEs allowed us to get direct feedback on the process and the tool.

Different formats (e.g., narrative, written, interview clips) make the resulting system more attractive. Videotaped interviews are a great way to get lively content, as well as a wide variety of content. People enjoy sharing their experience and being heard, particularly when it comes at low cost or effort for them (e.g., through blitz interviews). In addition, even we as concept providers were struck at how much more memorable are the video clips with the voice and facial expressions of the content providers, than their transcribed counterparts.

Even though the general concept is the same, it needs to be tailored and instantiated to the specifics of an organization. This goes for the structure of potential content (e.g., what attributes are to be collected for practices or evidence records), handling processes (shortcuts, quality gates/checklists), and even selecting tool support that integrates well with the culture of the organization or tools already in place.

Distributed knowledge handling cannot be done without tool support. Although EMPEROR is not tied to a specific tool, basic requirements for any tool supporting the method include distributed access, versioning history, and basic workflow support.

Having an explicit CM role to oversee routing of content is important as it ensures the efficiency of the processes by making sure that content is assigned to the SME that can best handle it and that each piece of content is handled by the SME who is best suited to handle it.

FUTURE RESEARCH NEEDED IN THE AREA

The implementation of the EMPEROR method in BPCh showed the feasibility of the approach and yielded extensive insights into how to set up and run a collaborative experience management system, but several issues remain that are worth further investigation.

Maintenance was only marginally investigated as it was not the focus of the project. We know the basic triggers for maintenance, but the overall process might be more streamlined. However, as content becomes older, maintenance procedures will have to be refined. Some aspects may be completely automated by tool support, such as triggering regularly scheduled maintenance. Further issues to consider deal with retiring content—when and how would that happen?

Any method like EMPEROR relies heavily on contributions. Therefore, it is worth investigating what motivates and continues to motivate content providers and SMEs to participate as actors in EMPEROR and how to keep them active. Although EMPEROR has much more hierarchical structures, several of Rafaeli and Arieli's conclusions on why people contribute to systems like Wikipedia (Rafael et al., 2008) can be transferred to the situation in EMPEROR.

Currently, practice summaries are mostly written "manually" by the SME, who compares and compiles existing evidence records for a practice. Because it is a tedious job to manually determine the "average" effect of a practice across several evidence records, tool automation or partial automation would alleviate the SME of a tedious task.

Determining optimal tool support requires close cooperation with experienced SMEs.

We believe that a system such as EMPEROR is a promising approach to gather and aggregate important, yet distributed information into high-quality content. Similar systems for aggregating and vetting experience would allow users to make informed decisions and would be helpful in many technical domains.

REFERENCES

Basili, Victor, et al. The software engineering laboratory: An operational software experience factory (Conference). ICSE. New York: ACM, 1992. Vol. 14.

Basili, Victor R., Caldiera, G., and Rombach, H.D. The Experience Factory. In *Encyclopedia of Software Engineering*, Vol. 1, pp. 469–476. John J. Marciniak (Ed.). New York: Wiley, 1994.

Becker-Kornstaedt, Ulrike C. Descriptive Software Process Modeling—How to Deal with Sensitive Process Information. *Empirical Software Engineering*, 6(December), 4, 2001.

The Cochrane Collaboration. 10 Key Principles. May 2011. http://www.Cochrane.org/about-us/our-principles.

Dangle, Kathleen, et al. Introducing the Department of Defense Acquisition Best Practices Clearinghouse. *CrossTalk*, May, 2005.

The Data and Analysis Center for Software Defense Authorization Act 2003. DACS. June 2011. https://www.thedacs.com/topics/BestPractices/section804.php.

DAU, The Defense Acquisition University. May 2011. www.dau.mil.

Houdek, Frank, Schneider, Kurt, and Wiesner, Eva. Establishing experience factories at Daimler-Benz: An experience report. ICSE. Washington, DC: IEEE Computer Society, 199Microsoft Sharepoint. Microsoft Corporation. http://www.microsoft.com/industry/government/products/Sharepoint2010/default.aspx.

Mongoose Technology. The 12 Principles of Collaboration.™ Guidelines for Designing Internet Services That Support Group Collaboration. White Paper. May 9, 2011. https://community.jivesoftware.com/docs/DOC-8202.

Office of the Secretary of Defense. A New Approach for Delivering Information Technology Capabilities in the Department of Defense. Report to Congress. Office of the Secretary of Defense, November 2010. http://dcmo.defense.gov/documents/OSD%20 13744-10%20-%20804%20Report%20to%20Congress%20.pdf.

Pinch, Trevor, and Kesler, Filip. How Aunt Ammy Gets Her Free Lunch: A Study of the Top-Thousand Customer Reviewers at Amazon.com. 2011. http://www.freelunch.me/home.

Pollard, Dave. Seven Principles of Social Networking: How to Save the World. July 25, 2005. http://howtosavetheworld.ca/2005/07/14/.

Rafael, Sheizaf, and Ariel, Yaron. Online Motivational Factors: Incentives for Participation and Contribution in Wikipedia. *Psychological Aspects of Cyberspace: Theory, Research, Applications*. Barak, A. (ed.) Cambridge, UK: Cambridge University Press, 2008. http://gsb.haifa.ac.il/~sheizaf/cyberpsych/11-Rafaeli&Ariel.pdf.

Seigenthaler and Wikipedia. Journalism.org. July 2011. http://www.journalism.org/node/1675.

Shaw, M.A., Feldmann, R.L., and Shull, F. Decision Support with EMPEROR. *Proceedings of the First International Symposium on Empirical Software Engineering and Measurement (ESEM07)*, poster track. Madrid, Spain: IEEE Computer Society, 2007. p. 495.

Shull, F., Feldmann, R., and Shaw, M. Building Decision Support in an Imperfect World. In *Proceedings of the Fifth ACM-IEEE International Symposium on Empirical Software Engineering (ISESE '06)*, Vol. II. Rio de Janeiro, Brazil: ACM-IEEE, 2006. pp. 33–35.

Shull, Forrest, and Turner, Richard. An Empirical Approach to Best Practice Identification and Selection: The U.S. Department of Defense Acquisition Best Practices Clearinghouse. International Symposium on Empirical Software Engineering (ISESE 2005). Noosa Heads, Australia: Computer Society, 2005. pp. 133–140.

Wikipedia, The Free Encyclopedia. 2011. http://en.wikipedia.org/wiki/Main_Page.

7

Real-Time Knowledge Management: Providing the Knowledge Just-In-Time

Moria Levy

INTRODUCTION

Knowledge workers are one of the most important assets to be managed. More employees in all organizations are becoming knowledge workers— workers whose knowledge is meaningful to their success. Productivity of these people is likely to be the center of managing people (Drucker, 1999).

Knowledge management (KM) deals with the productivity of the knowledge worker. Real-time knowledge management is a subset of KM focusing on knowledge that has to serve the knowledge worker in real time. The aim of this research is to investigate KM in conditions of real-time needs. The research assumption is that nonstandard KM solutions will be adapted for real-time needs.

The research took place between the years 2010 and 2011. As KM is a developing discipline, it is believed that the timing of the research has affected its results. Some KM solutions demonstrated were rather new (especially in the healthcare services sector) and did not exist 3 years beforehand. In 5 years, namely 2015 and further, the results may differ again.

This chapter is based on the following structure. First, real-time knowledge management is defined and described. It is assumed that real time requires customized solutions. As it is assumed that the expertise level of the role of the knowledge worker should affect the usage of KM systems, the expertise level of the knowledge workers should be taken into consideration. The next paragraph enlarges the discussion regarding the expertise dimension, and how different levels of expertise roles may use specific

KM solutions. Based on these two factors, regular versus real-time KM, and various KM expertise levels, the scope of the research is defined and the research methodology is explained.

For each role, defined as a real-time knowledge worker role, and representing a different level of expertise, the KM solutions are reviewed, as learned both from literature and from the field research conducted. This is repeated for each role in the scope of the research: service centers, banks, and medical physicians. Next, typical standard KM solutions are reviewed, enabling the comparison between the different KM real-time roles and non-real-time roles.

The research method chosen was grounded theory. Based on the findings, a theory and triggered architecture are suggested. The research is concluded by specifying theoretical and business implications and pointing out directions for further research.

REAL-TIME KNOWLEDGE MANAGEMENT

Knowledge workers, as any other type of workers, operate in an environment where spare time is always scarce. However, there is a difference between an engineer sitting behind his or her desk and trying to decide how to design a new electric circle, and a doctor who is speaking with a patient as well as deciding what checkups are indicated for that particular patient.

Gartner defines the real-time enterprise as "getting the right information to the right people at the right time" (Gartner, Inc., 2002). This definition is suggested by Kerschberg and Jeong (2005, p. 1) for just-in-time knowledge management: "the concept of Just-In-Time Knowledge Management is appealing in that, the goal is to provide the right information, to the right people, at the right time—just in time—so they can take action based on that information."

Malhotra (2005), based on Lindorff (2002), Lindquist (2003), Margulius (2002), Meyer (2002), Siegele (2002), and Stewart (2000), added to this definition: "without latency or delay."

Reviewing various sources, the two terms *real-time knowledge management* and *just-in-time knowledge management* are used by researchers alternately. Davenport and Glaser (2002) prefer the term *just-in-time*, while El, Omar, and Majchrzak (2004) and Mellor Gilhardi (1997) choose to use the term *real-time*. Some researchers focus on delivering the knowledge in real-time/just-in-time (i.e., Davenport and Glaser, 2002), without examining how critical it is to the specific user, while others (i.e., Kerschberg and Jeong, 2005) focus on the specific role where the just-in-time/real-time, is required due to the type of interface between the employee and the customer.

In this research, the term *real-time* was chosen. The focus of the research resides on those employees who have to decide and give answers here and now. These situations of real-time knowledge management differ from other job situations, as the employee has to respond to a person sitting in front of him or her, or speaking with him or her on the phone, waiting for a professional answer on the spot. Under these circumstances, the research aims to examine what types of KM solutions will suit, enabling the employee to best perform his or her job.

THE EXPERTISE DIMENSION

Examining the real-time factor by itself is not sufficient, as not all real-time solutions are similar. The usage of KM systems differs not only by the time one may have before making a decision, but it also may depend on the need for system assistance. This factor representing the need level was examined by comparing different groups of employees: on one side of the scale are the call center representatives, who are knowledge intermediates—they are hired with no specific knowledge, trained for several weeks, and their turnover is about a year. On the other side are the physicians/doctors, who studied for at least 7 years and in most cases stay in their profession for a lifetime. In between are front-line bankers and nurses. All of the above operate in conditions where the information based on a decision has to be provided in real time. The research assumption is that this different level of self- knowledge, and therefore the different level of need for the knowledge, will affect the KM solutions provided for the different roles.

RESEARCH SCOPE

The research examines the KM solution for real-time needs by examining the following knowledge worker groups. As shown in Table 7.1, the expertise level is based on average years of learning and typical turnover in the job.

TABLE 7.1

Researched Roles

Knowledge Workers' Group	Real-Time Type of Work	Expertise Level
Physicians (doctors)	Real time	Very high
Nurses	Real time	Medium-high
Front bankers (investment consultants, tellers)	Real time	Medium-high
Call center representatives	Real time	Low[a]
Back office employees (engineers, etc.)	Non-real time	Varying

[a] There are some sectors in which call center representatives require higher expertise level (like in computing and engineering problem-solving centers); however, this is not the majority, and these were not the examined cases.

Note: All real-time knowledge workers are in charge also for non-real-time processes. For example, a physician is in charge of checking results of patients' checkups in order to decide whether to proactively recommend a meeting with a patient. The research focuses only on the real-time processes of theses knowledge workers. In addition, the research focuses only on the usage of knowledge in real-time situations, and not on the creation of new knowledge in these situations.

RESEARCH METHODOLOGY

The research described in this chapter is based on the grounded theory methodology. The research questions the way KM is implemented for roles that are heavily based on real-time situations. In order to fully understand the issue and suggest an effective architecture for future businesses dealing with real-time roles, a grounded theory (Glaser and Strauss, 1967) qualitative research was conducted. Two methods were used for collecting the data: surveying organizations that include heavily based real-time KM roles, and reviewing the literature regarding published articles, describing real-time or just-in-time KM research.

The data was coded, sorted, and categorized. The categorization into groups was the key for explaining why different types of KM solutions were demonstrated for real-time situations. Categorizing real-time roles by level of expertise of the people filling in these roles gave the basis for

TABLE 7.2

Surveyed Organizations

Organization Type	Business Role	Number of Organizations Surveyed
Government	Call center representatives	1
Car industry services	Call center representatives	1
Telecommunications	Call center representatives	3
Banks	Account bankers (tellers)	2
Banks	Investment consultants	2
Healthcare services	Physicians (doctors)	2
Healthcare services	Nurses contact center representatives	2
Healthcare services	Administrative contact center representatives	1
Total		14 cases

the grounded theory, enabling the suggestion of a theory and, hence, an architecture for real-time business roles.

Here are the details of the data sources described with its categorization.

Survey Data Sources: A survey conducted in Israel that questioned knowledge managers in charge of providing KM solutions for real-time-oriented roles. Fourteen cases were examined, representing the largest organizations of each type in the country (see Table 7.2).

Literature Data Sources: A literature review, learning what KM real-time (and KM just-in-time) articles exist, and what solutions they describe. The literature represents worldwide knowledge reported cases.

The following paragraphs describe the KM solutions, demonstrated for the three main real-time business roles: call center representatives at service centers, front-line bankers, and physicians. These three roles are heavily based on real-time situations. In order to better understand the uniqueness of their solutions, we will first describe typical KM solutions for other business roles, where the work is not real-time oriented.

KNOWLEDGE MANAGEMENT (KM) SOLUTIONS FOR NON-REAL-TIME KM ROLES

KM solutions existed from the 1990s, serving various types of knowledge workers. As knowledge workers differ, one from another, there is no one system that fits all (Davenport, 2005). Among the different KM solutions known and described, there are portals, communities of practice, expertise locators, social networks, blogs, lessons learned systems, document systems, Web content management systems, wikis, and others. Yet, with regard to 2011, it was learned from articles as well as interviews of organizations in Israel that there are two commonly used KM systems: portals (intranets) and document repositories (whether as a network drive, a Web site including documents, or a formal ECM [enterprise content management] system). All other types of KM systems exist, yet, in most cases, do not serve as the main solution. The document repositories, most naturally, handle documents. The portals mainly handle documents, as well, added with some lists (bulletin boards, contacts, discussion groups, etc.). Most knowledge workers, when accessing some KM system, retrieve a document.

KM FOR CALL CENTER REPRESENTATIVES AT SERVICE CENTERS

"Call Centers are high-pressure work environments characterized by constant routine, scripting, computer-based monitoring, and intensive performance targets" (Houlihan, 2000). This is true, both for call centers, as well as for other types of service centers. The representatives at these organizations work mainly in real-time situations, where they are required to pass information to a customer as fast and yet professionally as possible. At these service centers, and specifically in the cellular sector, one of the first and largest sectors that has built service centers, employees pass organizational information to customers, through representatives who did not create the knowledge. In most cases, and specifically in those researched, this type of knowledge worker can be referred to as a knowledge intermediate; the worker does not possess the knowledge, rather, he or she transfers it from the organization to the customers. These knowledge workers do not need specific preacademic education in order to qualify for the job. The training is provided on-site, and usually the training duration is 2 to 3 weeks.

Davenport, in his book *Thinking for a Living* (2005), refers to call center representatives, and the applications that they use, as knowledge workers. "The applications for Call Centers include Customer Relationship

Management tools, tools for scripting conversations with customers, knowledge tools for solving customer problems, and tools for capturing customer feedback" (Davenport, 2005, p. 106).

The findings from the data sources, articles (see Table 7.3), as well as interviews, conclude the same results: medium and large service centers are automated. The representatives use unique application systems: The information and knowledge are structured and displayed as knowledge items. The knowledge items are structured, and each group of items shares the same structure. This is different than documents that are described as the main knowledge item for non-real-time workers, and where the majority of the documents are designed in free format. The research also investigated the nature of these knowledge items. In the interviews, the knowledge solutions were asked as to the knowledge item's nature: Where are they stored? How are they designed? How are they accessed? The findings teach that the knowledge items are stored in unique knowledge bases, named *Contact Center Applications* or *Knowledge Management Applications*. The systems resemble Web content management systems as they are organized for handling structured knowledge items and are suited for the vertical service center's specific needs. There are two typical ways to display the knowledge. One is by using scenarios, also named scripts, guiding the representative as to what to ask, say, or in some cases, how to act. This type of knowledge item is mentioned by Davenport (2005) where he describes the British Telecom Call Center case study. The second way is by knowledge items structured in sets of fields, while each representative decides what subset of fields and, hence, what information to use in each case. The issue of how to organize the knowledge items is important, as it affects the response time of the call center representative.

In most interviewed organizations, a mixture of the two methods was found, rather the balance (how many scripts, how many structured fields) varied from organization to organization as shown in Table 7.4 (data displayed in percentages).

Speaking with the knowledge managers in charge, no one answer was agreed upon as to whether scripts are preferred or structured fields. No organization demonstrated more scripts than structured fields, yet some organizations prefer scripts wherever possible, while others use it only in specific places. In one organization, even though they used 90:10 in favor of fields, it was stated that scripts are better, and they are now in the process of changing structured fields to scripts in order to have more

TABLE 7.3

Reviewed Articles

Article	Organization Type Described	Business Roles Described
Knowledge Management for Call Centers	(General)	Call center representatives
Contact Center Knowledge Management	(General)	Call center representatives
The Application of Knowledge Management at Call Centers	(General)	Call center representatives
A Simulation Approach to Restructuring Call Centers	Utilities	Call center representatives (tech support for a wide range of products and services)
Knowledge Management at Call Centers	Car industry services	Call center representatives
Critical Issues in Research on Real-Time Knowledge Management at Enterprises	(General) Hardware	(General) Call center representatives, customer support
Eyes Wide Shut? Querying the Depth of Call Center Learning	(General) Insurance	Call center representatives
Getting to "Real-Time" Knowledge Management: From Knowledge Management to Knowledge Generation	Utilities	Consultants
Knowledge Management at Malaysian Banks: A New Paradigm	Banks	(General)
Accelerating Customer-Oriented Banking with Knowledge Management	Banks	Front-line (branches and sales offices)
Knowledge Management in Banking Industries: Uses and Opportunities	Banks	(General)
Evaluating the Efficacy of Knowledge Management Toward Healthcare Enterprise Modeling	Healthcare services	(General)
Knowledge Management in Evidence-Based Medical Practice: Does the Patient Matter?	Healthcare services	Physicians (doctors)

TABLE 7.3 (CONTINUED)

Reviewed Articles

Article	Organization Type Described	Business Roles Described
Look Before You Leap: Learning from the Experience of a Flagging KM Initiative at a Healthcare Organization in Asia	Healthcare services	Physicians (doctors), nurses
Just-In-Time Delivery Comes to Knowledge Management	Healthcare services	Physicians (doctors)
Total		15 articles

Sources: Data from Hafizi and Nor, 2006; Jayasundara, 2008; Lam and Law, 2004; Rasooli, 2006.

TABLE 7.4

Structuring Methods Usage

Organization	A	B	C	D	E
Structured fields	70%	90%	60%	70%	80%
Scripts	30%	10%	40%	30%	20%

Note: A, B, C, D, E represent organizations that were interviewed.

scripts. In some other organizations it was stated that the representatives prefer information displayed as fields, leaving them the independence to decide in what order to access the information. Another organization stated it holds both, scripts for new representatives, and structure fields for the others.

As to the large numbers of representatives at call centers, and as to the high effect, as stated, of the knowledge organization on the ability to respond efficiently, most interviewed organizations, and surely all in the cellular sector, continually work on improving the knowledge trees (menus) and refining functionality of the search engine, adding facilities to effectively find the requested knowledge as simply and fast as possible. These organizations have unique teams in charge of writing, preparing, and structuring the knowledge properly (see Table 7.5).

All interviewed organizations were satisfied with having a unique KM system.

It can be concluded that for this role the KM system is unique and planned, focused on its real-time business role orientation.

TABLE 7.5

Real-Time Knowledge Management (KM) System Attributes

Topic	Real-Time KM System	Non-Real-Time Typical KM System
Entity managed	Knowledge item	Document
Structuring level	High	Low
Accessibility level	High/embedded	Varying/mainly stand-alone

KM FOR FRONT-LINE BANKERS

Front-line bankers may be defined as employees working in the banking sector who serve the customers of the bank, as opposed to back office bankers who provide services within the bank to other employees, but not mainly directly to the end-customer holding an account at the bank. Analyzing the data sources for this sector was not trivial. Not all articles distinguished what type of knowledge worker at the bank KM systems are serving. KM solutions described in articles included mainly document systems (libraries, maps pointing to documents, etc.) and intranets (some being part of communities of practice).

The interviews held distinguished among the front-line bankers two types of knowledge workers: investment consultants and account managers (tellers). These two types are front-line knowledge workers who both need preacademic education in order to qualify for the job. Yet, the investment consultant has more years of experience, and the job requires higher levels of training.

The largest two banks in Israel were interviewed, resulting in the following:

Both banks have built highly invested KM solutions for investment consultants. Yet, these all fall into the category of portals/intranets.

For account managers, the two banks offered different solutions: One bank offered an Intranet. Yet, the second realized that in order to make these front-line knowledge workers more effective, a well-structured knowledge base is required. They launched, 6 months before the interview took place, a KM system based on structured knowledge items, mainly structured as scenarios, accompanied by documents and structured fields. Their users were very satisfied with the change that eased knowledge usage. Special care was put into improving accessibility and easing the method of finding the relevant knowledge item, both through routine work on the menus (knowledge trees) as well as on investing in purchasing an advanced search engine. This KM solution, resembling KM solutions demonstrated at call centers, is unique and was not experienced in other medium to large banks in Israel. It should be noted that the need is emerging. Also, a third medium to large-sized bank interviewed on this specific issue only, stated that the front-line bankers have direct lines to the call center representatives who use such a system, and they use them to answer customer queries and quicken response times. Additionally, the bank that offered the structured KM system was in its last stages of implementing another system for all bankers, where when right clicking a term, in all systems, a window would pop up with the initial, most relevant knowledge for that term. The knowledge presented was structured.

The conclusion is that in most cases, these real-time knowledge workers work with the same KM systems as the typical non-real-time knowledge worker. There are some cases in which unique systems that resemble the structured knowledge bases were built to ease and accelerate response, yet it is not clear if these are and will be an exception or if they forecast a new trend for the future.

KM FOR MEDICAL PHYSICIANS

Davenport and Glaser (2002) describe the context in which medical physicians work: "Dr...has a big problem, one common to all Physicians. There is so much knowledge available about the work, that he cannot possibly absorb it all. He needs to know something about almost 10,000 different diseases and symptoms, 3,000 medications, 1,100 laboratory tests, and many of the 400,000 article" (Davenport and Glaser, 2002, p. 107). Medical physicians may be viewed as extreme knowledge workers in terms of education, experience, information and knowledge relevant to the job, and importance of the decisions taken: Initial learning includes at least 7 years of education and additional years of expertise; the relevant knowledge bases that can help them perform in their job are huge and grow every year; and the importance of making the right decisions is critical. Furthermore, as medical physicians are real-time knowledge workers, they have to make many of their decisions fast, while they interact with a patient.

Several articles were reviewed in order to learn what KM solutions are offered to medical physicians: Cheah and Abidi (1999) describe a healthcare enterprise organizational memory knowledge base that is divided into sub-knowledge bases. These enable access to protocols (policies and procedures) separately for every unit. Types of units may include, for example, outpatient department, emergency unit, and dental clinic. Additionally, a general knowledge base includes medical procedures/treatments and best practices (Cheah and Abidi, 1999). This type of KM solution is based on documents and represented via intranets, help desks, workflow, groupware, document management, and so forth. Interrelations exist between the subknowledge base, in means of updating one, as triggered for others being updated.

A similar KM solution, yet less described in detail is presented by Chua and Goh (2008), regarding a case study of a healthcare organization. The solution was designed, as in our interest, for real-time knowledge workers (doctors and qualified healthcare professionals) based as a Web system (intranet). Another example of a KM solution for making clinical decisions is described by Boateng (2010) and is based on evidence-based medical practice. This solution is also based, as the former described ones, on a search in the knowledgebase for relevant articles; hence, it is based on documents (Boateng, 2010).

Davenport and Glaser (2002) describe a different type of KM solution for medical physicians. While the medical physicians log on into their operational systems and work within the patient's record, the KM system operates in the background. When ordering a drug, for example, the system checks if there may be any allergic reactions to any other medications prescribed. Recommendations are also offered in cases of ordering tests, based on the recorded systems. This type of KM solution is based on business rules, which are a unique type of structured knowledge item. "The power of (this type) of knowledge-based order entry, referral, computerized medical-record, and event-detection systems is that they operate in real-time" (Davenport and Glaser, 2002, p. 110).

In Israel, interviews were conducted at the two largest healthcare service providers in the country. In both cases, the knowledge was embedded into the operational systems: In one organization, a business rule system was implemented, as described in Davenport and Glaser (2002). This system had been implemented for 2 years, replacing a former KM solution based only on an intranet. At the other organization, the knowledge embedding solution enabled quick access to drugs and tests knowledge bases directly from the user's record, directing the doctor directly to the drug or test in context when right clicking the term. The solution was new and yet tested as a pilot (implemented for a few months). Initial responses were positive. It should be noted that these two solutions are similar, yet different; they both embed the knowledge into the operational system, yet while the first solution presented works in "push" mode, the second is "pull" oriented. For non-real-time KM roles, the usage of business rule systems, as well as other embedded KM systems in operational systems can be found, yet is rather rare.

The conclusion drawn is that this sector is probably experiencing a change toward the preferred solution for real-time KM. From intranets and document management systems, which are stand-alone KM systems based on documents, the organizations are moving toward embedded KM systems, offering their knowledge workers just-in-time recommendations based on structured knowledge items in pull or push mode.

SUMMARY

In this research, it was found that real-time knowledge workers use KM systems. Findings suggest that only in the call center sector, KM unique systems are the default; yet all types of real-time roles have been found to experience useful real-time unique KM systems. There appears to be a positive trend of enlarging these systems over time (Levy, 2009).

Even though unique KM systems are typical for some roles more than others, based on the self-knowledge of the type of knowledge worker holding the position, all roles benefit from the unique systems provided, and the systems can be used in real-time situations. Wherever previous nonunique KM systems existed, organizations were satisfied with the change. No organization was found to withdraw a unique KM system after being built, returning back to an intranet or documenting system.

Also, similar types of real-time unique systems were found across the different roles.

SUGGESTED ARCHITECTURE

Architecture for providing KM solutions for real-time-oriented roles and for other real-time scenarios is suggested by emphasizing accessibility of the KM solution.

Based on the research, a theory is suggested: All real-time knowledge workers can benefit from working with unique KM systems and should work with such systems.

As KM unique, real-time systems were found to be more costly, in terms of content processing and maintenance, the main criterion defining where to invest in building these systems is cost-effectiveness (i.e., where the organization benefits more).

Typical examples of these may include holding fewer knowledge workers for the job, as in the case of call representatives; benefiting from less errors in decisions, like in the case of physicians; benefiting from more professional responses to customer queries; and so forth.

In order to suggest a suitable recommendation, the real-time KM system has to be defined.

A real-time KM system is a KM system that benefits from eased accessibility, both externally and internally, as described below:

External Accessibility: Easing the path of the user to the knowledge item.
Internal Accessibility: Easing the readability and hence the understanding of the knowledge, once it was reached (Levy, 2009).

Real-time KM systems differ from regular KM systems in the following ways:
External accessibility can be achieved in a few ways:

"Pushing" the knowledge to the real-time knowledge worker, while he or she uses an operational information technology (IT) system
Easing the "pull" of the knowledge by enabling access using right click, or any similar technique
Enabling direct access to the knowledge items through search engines and fine-tuned menus (knowledge trees)

External accessibility is demonstrated in Figure 7.1.

Note: Dealing with external accessibility is not unique to the real-time KM systems. However, more care and emphasis were found for real-time roles, improving both the hierarchical navigation through the knowledge trees (mainly manually) and the search mechanism (mostly automatically, improving the software involved). In non-real-time-oriented roles, only few usages of business rules and other push mode mechanisms were found.

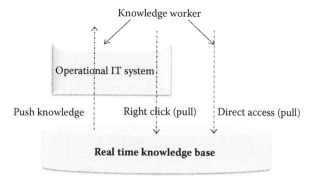

FIGURE 7.1
External knowledge access.

KM found practices as follows:

- Wherever operational IT systems exist, it is preferred to use these to serve as the interface to the real-time knowledge.
- Access, and especially pull-oriented access, should be enabled through several access channels, wherever possible.

Internal accessibility can be achieved by using structured knowledge items, rather than documents. Three types of knowledge items can be used:

1. Business rules: Short "bottom lines" advising the knowledge worker what to do.

   Usable when relevant: Knowledge is clear, short, and precise.

2. Structured fields: May be embedded in subgroups.

   Usable when the body of knowledge is larger, enabling the knowledge worker to decide what specific knowledge is more suitable for the specific case; usable also when there is no unique answer, the knowledge serving decision support.
   Demonstration: See Figure 7.2.

3. Scripts: Usable for new users; usable when there is a known workflow for the knowledge decision.

   Demonstration: See Figure 7.3.

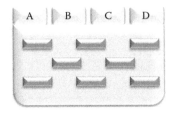

FIGURE 7.2
Internal knowledge access—structured fields.

FIGURE 7.3
Internal knowledge access—scripts.

KM found practices as follows:

- Simplicity is most important for easing understanding and usage of content. There is no clear answer to the question whether short (as in structured items) or lengthy (as in scripts) is simpler, and the decision is to be taken in each organization as to its context, striving to maximize simplicity.
- The employees in charge of writing the context may be part of each professional unit in charge of the content, or in a centralized unit. It was found that defining the internal structure and writing the knowledge items is a profession, and it is preferred to organize these technical writers as one centralized unit.

Organizations should strive to design for their real-time knowledge workers' KM systems, including improved external and internal accessibility, in one or more of the defined types above. It was not proven that real-time knowledge workers will not use and benefit from other types of KM systems, yet, it was found that organizations that turned to using these types of unique KM systems were satisfied with the change and none wanted to return to classical KM systems (such as intranets or document systems), even though these systems are expensive to build and expensive to maintain.

CONCLUSIONS

Drawing from the research, possible conclusions as to the need of unique KM solutions for real-time scenarios are as follows:

Unique KM systems were found in usage, serving real-time-oriented knowledge workers. Different real-time role types were found to use these systems at different levels and in typical ways, yet unique KM systems were in use at all examined role types.

It is recommended to integrate all the various solution types found in usage of the different roles and to implement unique KM solutions based on these for real-time roles in the organization. These solutions will be unique in the way they ease both external access and internal access to the knowledge worker. Organizations should prioritize implementation of these solutions, as they are more expensive to implement. Thus, roles where knowledge is more in need should be prioritized (as for lower-level expertise knowledge workers, scenarios in which the organization considers the decisions to be more critical, etc.).

FURTHER RESEARCH

The research was limited. The research scope was mainly based on three sectors of real-time-oriented users. It is recommended to further investigate other sectors as well, validating the findings. The scope was limited in the amount of organizations examined. And finally, while the research took place, a change was viewed; organizations are starting to implement real-time KM solutions. It is recommended to reanalyze and conduct a postaudit of the real-time KM topic in these organizations and others after several years of KM usage.

REFERENCES

Boateng, W. (2010). Knowledge management in evidence-based medical practice: Does the patient matter? *Electronic Journal of Knowledge Management,* 8(3): 281–292.

Cheah, Y.N., and Abidi, S.S.R (1999). Evaluating the efficacy of knowledge management towards healthcare enterprise modelling, International Joint Conference on Artificial Intelligence IJCAI'99, Stockholm.

Chua, A.Y.K., and Goh, D.H. (2008). Look before you leap: Learning from the experience of a flagging KM initiative at a healthcare organization in Asia, *Aslib Proceedings: New Information Perspectives,* 60(4): 335–348.

Davenport, T.H. (2005). *Thinking for a Living,* Harvard Business School, Boston, MA.

Davenport, T.H., and Glaser, J. (2002). Just in time delivery comes to knowledge management, *Harvard Business Review,* July. pp. 107 and 110.

Drucker, P.F. (1999). *Management Challenges for the 21st century,* HarperCollins, New York.

El, S., Omar, A., and Majchrzak, A. (2004). Critical issues in research on real time knowledge management in enterprises, *Journal of Knowledge Management,* 8(4): 21–37.

Gartner, Inc. (2002). *The Real-Time Enterprise,* retrieved from http://www.gartner.com/1_researchanalysis/focus/ebus113001.html

Glaser, B.G., and Strauss, A. (1967). *The Discovery of Grounded Theory: Strategies for Qualitative Research,* Aldine Transaction, Piscataway, NJ.

Hafizi, M.A., and Nor, A. (2006). Knowledge management in Malaysian banks: A new paradigm, *Journal of Knowledge Management Practice,* 7(3). http://www.tlainc.com/artcl120.htm

Houlihan, M. (2000). Eyes wide shut? Querying the depth of call centre learning, *Journal of European Industrial Training,* 24(2/3/4): 228–240.

Jayasundara, Chaminda C. (2008). Knowledge management in banking industries: Uses and opportunities, *Journal of the University Librarians Association of Sri Lanka,* 12: 68–84.

Kerschberg, L., and Jeong, H. (2005). Just-in-time knowledge management, keynote in *Third Conference on Professional Knowledge Management,* Kaiserslautern, Germany.

Lam, K., and Law, R.S.M. (2004). A simulation approach to restructuring call centers, *Business Process Management Journal,* 10(4): 481–494.

Levy, M. (2009). Leveraging knowledge understanding in documents, *Electronic Journal of Knowledge Management Practice,* 7(3): pp. 341–351.

Lindorff, D. (2002). Case Study: General Electric and real-time, *CIO Insight,* November 11. Retrieved from http://www.cioinsight.com/article2/0,3959,686147,00.asp

Lindquist, C. (2003). What time is real time? *CIO Magazine,* February 10. Retrieved from http://www.cio.com/online/techtacL021003.html

Malhotra, Y. (2005). Integrating knowledge management technologies in organizational business processes: Getting real time enterprises to deliver real business performance, *Journal of Knowledge Management,* 9(1): 7–28.

Margulius, D.L. (2002). Dawn of the real-time enterprise, *InfoWorld,* January 17. Retrieved from http://www.infoworld.com/article/02/01/17/020121fetca_1.html

Mellor Ghilardi, F.J. (1997). Getting to "real-time" knowledge management: From knowledge management to knowledge generation, *Medford,* 21(5): 99–101.

Meyer, C. (2002). Expert Voice: Christopher Meyer on the accelerating enterprise, *CIO Insight,* November 2. Retrieved from http://www.cioinsight.com/article2/0,3959,675333,00.asp

Rasooli, P. (2006). *Knowledge management in call centers,* Master thesis, Department of Business Administration and Social Sciences, Lulea University of Technology, Sweden.

Realcom. (2009). Accelerating customer-oriented banking with knowledge management. Retrieved from http://www.realcom.co.jp/en/case_studies.html

Siegele, L. (2002). The real-time economy: How about now? *CFO (The Economist)*, February. Retrieved from http://www.cfo.com/printarticle/0,5317,6651%7C,00.html

Stewart, T.A. (2000). How Cisco and Alcoa make real time work, *Fortune*, May 29.

8

Building Vertical and Horizontal Networks to Support Organizational Business

Maureen Hammer and Katherine Clark

INTRODUCTION

Historically, the Virginia Department of Transportation (VDOT), a large, mature organization, could count on an unending pool of employees who slowly worked their way through and up the organization. Managers determined what resources they needed to do the job and got the employees they requested with the associated knowledge and experience they felt was necessary. Staff were hired at entry-level positions and worked their way up through the organization. What happens when that organization goes from over 12,000 employees down to 7,500 through purposeful downsizing to accommodate a new business model but which puts at risk the knowledge and ability to deliver the program with which it is charged? While new employees to the organization may come in with education and some work experience, they will not have the institutional knowledge or access to the hidden networks that were the basis of success for more than 100 years. How do you address those issues and ensure that you are prepared for additional downsizing and changes?

BACKGROUND ON THE VIRGINIA DEPARTMENT OF TRANSPORTATION (VDOT)

The Virginia Department of Transportation (VDOT) was founded in 1906 as the Department of Highways to engineer and construct Virginia's primary roads, transforming horse and wagon tracks into a modern roadway network. In the 1930s they accepted local roads into the system and began to manage those as well. (Today, that totals almost 58,000 miles of road and more than 12,000 structures—bridges, tunnels, and ferries.) In the 1950s, this mission was expanded to successfully participate in the construction of the interstate highway system. Both of these endeavors were well served by a bureaucratic organizational structure that relied on each division and section doing their part and handing it off for the next step. Employees were brought in at entry-level positions and "apprenticed" into the organization, building knowledge, expertise, and networks as they went.

In the 1980s, the agency began to contract some of the work to private industry, whereas in the past it had performed it all in-house. In 1992, the agency, experiencing severe budget constraints, made an early-retirement option (ERO) available to staff in order to reduce operating costs related to salaries resulting in about a 7% reduction to 11,600 authorized positions. Three years later, in 1995, a new governor was elected to office determined to reduce state government size, and a Workforce Transition Action (WTA) was implemented, leading to the loss of 1,300 positions (slightly more than 10%) although the decision was made to refill 300 critical positions so that ultimately staff levels were reduced overall by about 10% to 10,357 authorized positions. Additional reductions in staff occurred bringing the agency down to 9,500 authorized positions in 2009. In 2010, the agency again downsized to a goal of 7,500. Today, approximately 15% of current employees are eligible for full retirement.

The authors conducted interviews with managers who had been present during these first two periods of staff reduction, both in the field offices (districts) and the central office supplemented by examination of documents, such as the agency newsletter, from this same period. Unattributed quotes in the following are from these interviewees.

Traditional Knowledge Transfer within the Agency

Traditionally staffing level supported redundancy of knowledge and people tended to stay in the agency for decades, sharing experiences throughout their career progression. The agency used the apprenticeship or military model. Employees learned from more experienced employees and slowly moved up the ladder with new assignments when management determined that they were ready. As they moved up the ladder they built personal networks with colleagues which supported them later.

You don't often hire somebody into a leadership position who didn't come from inside, that was the model. There was no way you were going to hire very many people from outside unless you hired them in an entry-level position. You'd put them in that job until they'd learn that and you would put them in another job until they'd learn that one, and then you'd move them to another and they [would] learn that job.

1991 Early Retirement Option (ERO)

Interviewees agreed that there was time to plan for the early buyout in 1992. The belief was that it was orderly and based on the economic needs of the state—reducing operating costs by reducing personnel costs. The agency could examine employee tenure levels and determine who was eligible and then have management make and implement plans to transition knowledge and assignments. Many managers perceived the reduction as an opportunity to reorganize departments more efficiently and to alter the focus of work as needed. Ultimately there was about 7% reduction in staffing.

The types of people taking this retirement were within 5 years of retirement, a great many of whom were in leadership positions. Staffing levels at the time meant that there was a second and third tier of employees who could step in to replace retirees in most cases; although there were a few locations that would take a hard hit. Knowledge had not necessarily been transferred, but the remaining employees had exposure to a lot of experiences and, therefore, to much of the expertise of those who had left which would support them going forward. On the whole, it was felt that the agency was able to manage around the knowledge loss.

What did get lost was the knowledge of how to manage the space, or relationships, between knowledge blocks. One interviewee described a process where knowledge "blocks" comprised specific components of the program. For example, if you are manufacturing cars, one block might be building the engine and another formation of the body parts and another creation of the electrical system. Between each of those blocks you have the connections, the knowledge of how to move from one block to the next and to put the components together to develop a successful end product. The people who left in 1992 were those with the experience to know how to make connections:

> You had this wave of institutional knowledge and keepers of the processes and how things "worked"—people who knew who to talk to, who to move things to, how to get stuff moved to all of these mid-managers. Mid-managers,

who really weren't ready to take up the mantle, really didn't know much about how their [part] of the world connected to the rest.

1995 Workforce Transition Act (WTA)

The downsizing in 1995 was different from the early retirement option in 1992. Where in 1992 the downsizing was perceived to have been done for financial reasons, in 1995 the perception was that the downsizing was more political and in response to the belief that government should be smaller. The prevailing attitude was that "the best kind of state employee is an employee not on the payroll." This resulted in a great deal of anger and fear among employees and more chose to take the offer this time. There were also different eligibility rules, the 1995 Workforce Transition Act (WTA) offered very favorable financial incentives to people to leave the agency—not only those close to retirement, but midcareer employees as well. Anyone wanting to participate was allowed to do so, as the director of the agency felt that if you told an employee who wanted to leave he or she had to stay, that employee would be angry and unproductive, which meant that there could be less than a month between the time the employee decided to take the option and the time the employee physically left. The WTA resulted in about a 10% reduction in employees. However, the agency leadership successfully argued that certain critical positions needed to be retained to allow the agency to continue its work. While the incumbent might take the buyout, the agency would be able to fill the position.

As stated by one person interviewed, there was a "loss of people in the prime of their careers, people with both expertise and experience." To continue with the analogy used earlier, if people who left in 1992 provided the continuity between the knowledge blocks, the people who left in 1995 were often the people who made sure that knowledge blocks worked successfully. We no longer knew what was in the block of knowledge that we had to have to perform the work. Whereas in 1991 we lost the ability to manage between blocks of knowledge, now we compromised or lost the blocks. Further, what did not become apparent until this buyout, was that we had done a poor job of documenting processes and procedures.

In 1995, we were losing people who had been trained and mentored by more experienced employees, and who had not yet completed mentoring and training new employees. We faced not knowing how to resolve problems because we lost the experience. It was also felt that this downsizing had a huge impact on continuity of leadership. With the WTA, the agency had lost almost 20% of both top and mid-level managers in less than 5 years.

2008 to 2010 Reduction in Force

In the 2008 to 2010 time frame, the agency went through another major downsizing, this time losing primarily newer and younger staff. These cuts were made in response to legislative mandates and much like in 1995 had an impact on employee morale. Ultimately, as a result of layoffs and attrition the agency was reduced to about 6,800 employees and is now hiring to attain the approved level of 7,500 employees to effectively deliver the increased construction program.

Impacts

The major difficulty as described by the interviewees who had been in the field and responsible for implementation of the program was that a gap in knowledge and experience was created following the reductions in the 1990s. Managers also had to address the anxiety and stress of remaining employees who knew they did not have the knowledge and who also recognized that despite the reduction in employees the work still had to be accomplished.

The perspective of those interviewed who had been at headquarters and involved in making agency policy was less stark, suggesting that this second round created a "blip" in the delivery of the program and actually was a positive thing in that it resulted in new leadership who were not married to the old way of doing things:

> You filled a lot of positions internally with very young folks and they had to learn on the fly, do the best they could do. In some respects that's not a bad thing. In other respects what it means is that they missed out on the apprenticeship process that you get when you have an opportunity to learn from someone who's mentoring you, as had always been done at VDOT; it was a very different time.

To perform this work it was necessary to negotiate cooperation between the different locations to get the necessary assistance:

> I would call the other districts and see if anybody wanted to help, or volunteered to help, or could offer to help…Yes, it was up to me and my staff to

prevail on the kindness of others. That worked a little bit but not for very long. I put together a significant analysis of our workload compared to other districts and made a plea to the [leadership] for some additional staff as well as some support through some contracts…eighteen months later; I got six positions to replace the seventeen that had left. We got some contractual help and some help from some other districts, but it was never very satisfactory.

Today

The program today is more affected by political changes, has a greater focus on bringing in new expertise from outside the agency, and requires a different perspective of knowledge sharing. We have seen the creation and solidifying of information "stovepipes" in which communication occurs in relatively narrow segments of the staff, but not across those segments. For example, while staff in one section may talk to each other across the state, they do not necessarily talk with staff in other sections across the state or even with staff across the hall but in a different functional area. It is no longer possible to count on an ample workforce building up a large store of knowledge and passing it on in an extended apprenticeship relationship over a long period of time. People left against their will, because they are part of a demographic large retirement wave, or because they were made offers that they could not refuse. Morale was seriously impacted, thereby making those offers even harder to refuse. At the same time, there was a development of internal leadership with a stronger business focus. The agency was both forced to (and had the opportunity to) bring in new expertise and to look at things in a different way than it had in the late 1990s and early 2000s. The ability to bring in new people to the agency was sharply curtailed with current freezes on hiring, or on hiring at market rates. Loss of staff in the last downsizing resulted in openings in the agency that are being filled by less experienced staff. In the long term this movement of people into new positions may well be positive, but in the short term it resulted in decreased experience in several important roles.

COLLABORATION AND NETWORKING PRINCIPLES APPLIED

What happens when large numbers of staff leave an organization and take their expertise and knowledge, their part of the organization's memory of how work is accomplished, with them? According to Lesser and Prusak (2001), when organizations downsize, it is often the most knowledgeable employees who leave first resulting in damaged critical social networks and an increase in time needed for knowledge transfers. These knowledgeable employees built and maintained extensive networks used to share knowledge. "What really distinguishes high performers from the rest of the pack is their ability to maintain and leverage personal networks. The most effective knowledge workers create and tap large, diversified networks that are rich is experience and span all organizational boundaries" (Cross et al., 2003, p. 20). This reflects what happened in our agency, so reestablishing networks between these knowledge workers to transfer knowledge, to create a culture that supports innovation and creativity, and to build solidarity was critical.

Goffee and Jones (1996) identify two underlying human relations in a community: sociability (the level of friendliness) and solidarity (the support and pursuit of shared goals). The culture can be networked (high sociability, low solidarity), mercenary (low sociability, high solidarity), fragmented (low sociability, low solidarity), or communal (high sociability, high solidarity). According to Smith and McKeen (2002), there are three underlying organizational attributes that facilitate a knowledge-sharing culture: high sociability and high solidarity, emphasis on fair processes and outcomes, and recognition of employees' work.

High sociability facilitates employees' willingness to seek knowledge from peers and to be a source to others:

> Benefits of a positive social interaction culture, with respect to knowledge sharing, include employees who are more knowledgeable about their colleagues' potential for being knowledge sources, as well as employees who trust more colleagues, and who trust them more completely, and who are willing to share knowledge with them as a result (Connelly and Kelloway, 2003, p. 295).

High solidarity ensures that employees share and support common goals that benefit the organization as a whole.

According to Davern (1997),

> A social network consists of a series of direct and indirect ties from one actor to a collection of others, whether the central actor is an individual person or an aggregation of individuals (e.g., a formal organization).

A network tie is defined as a relation or social bond between two interacting actors (p. 288).

Ties determine how closely related nodes are within the network. The stronger the tie between two actors, the more the ties held with others are similar, thus creating a redundancy in knowledge (Granovetter, 1973). He goes on to note that weak ties can be efficient for knowledge sharing because they give access to new information by bringing together groups and individuals in organizations that do not interact regularly (Granovetter, 1973). Weak ties also lead to innovation and creativity. However, weak ties can make it difficult to absorb complex, tacit knowledge (Augier and Vendele, 1999).

Knowledge transfer refers to identifying knowledge held by an individual or group and sharing that knowledge with another individual or group, resulting in a change of how the business process is approached, considered, or handled. "Although knowledge transfer in organizations involves transfer at the individual level, the problem of knowledge transfer in organizations transcends the individual level to include transfer at higher levels of analysis, such as the group, product line, department, or divisions" (Argote and Ingram, 2000, p. 151). Okhuysen and Eisenhardt (2002) argue that for groups to be effective, "knowledge that is 'owned' by individual members of such groups must spiral up to groups and even organizations, where it can be exploited to further the goals of the organization" (p. 370). The transfer of knowledge is evidenced when a change in performance of the organization results.

Many organizations support the sharing of knowledge through the use of communities of people interested in the same knowledge area or practice. These communities gather to share explicit and tacit knowledge within an organization or profession. "At work, communities of practice can exist solely within an organizational unit; they can cross divisional and geographical boundaries; and they can even span several different companies or organizations" (Burk, 2000, p. 18).

CASE STUDY

In large part due to downsizing and to the increased politicization of the agency, the culture that was in effect at VDOT when the Knowledge

Management Office was initiated in 2003 evidenced some important characteristics that seriously impacted collaboration and innovation.

- Functional silos were well-established, both geographically (nine districts and the central office) and organizationally (30 different divisions in the central office which had representatives in the districts).
- Documentation of policies and procedures was rudimentary as the agency depended largely on redundancy in the workforce as a means of passing on knowledge.
- The matrixed organization had shifting roles and responsibilities that changed as politics shifted.
- Executive leadership and decisions were politicized and frequently changing.
- Multiple, frequent audits resulted in a risk-minimizing compliance-based culture where innovative behavior was more likely to be punished than rewarded.
- Employee morale was poor.

Basing our program on the theories described earlier, the VDOT Knowledge Management Program implemented communities of practice (CoP) (see Figure 8.1) as a primary tool to support and promote collaboration and knowledge sharing across the silos that existed. Beginning with the idea that knowledge sharing is fundamental to collaboration, we chose to begin by establishing CoP with the intention of establishing networks that would be able to address current and evolving needs and thereby support efficiency and effectiveness in the workforce.

A Community of Practice

IS a group

– That has shared competence and expertise

– That shares knowledge, builds relationships, and learns

– That shares resources, experiences, tools, and responses

To address common issues or problems.

FIGURE 8.1
Definition of community of practice.

We formed approximately 50 CoP in the past 8 years with varying longevity—at any given time, approximately 30 are active. These are groups of people coming together to focus on a particular area or agency issue. The breakthrough is that the membership of these groups is not isolated within function or region, but integrates staff across geographic or organizational boundaries. In this way, CoP contributes to both lateral and vertical organizational change.

In breaking down communication stovepipes, management increasingly sees the CoP as a tool for use to reinforce and focus staff and to solve problems and improve processes by sharing information and knowledge across previously impenetrable lines. In addition, CoP provided a focused "voice" for staff to articulate information for management in a way that it can be used to review and impact policy. These communities also provided stability through establishing networks during a period of constant change by providing new employees with immediate networks and to long-term employees with expanded networks.

Communities of Practice (CoP) as a Network Creation Tool

As noted earlier, a CoP is a group with relevant expertise which is focused on a specific area or issue to achieve an outcome. At VDOT, KM's role in a CoP is to create a framework within which the members, the subject matter experts (SMEs), are able to focus on and accomplish specific objectives. KM does not develop the solutions, but we use a facilitation model that keeps pointing in the direction of producing an outcome. The outcome can range from specific tools, to improved understanding, to sharing information. Although not always measurable, we keep the idea of needing to stay focused on outcomes from these meetings in play as part of the discussion. CoP members are regularly asked if they feel that the CoPs are a good use of their time and energy. Their response is a resounding yes, and they indicate that the value is high for the organization.

In large part, this KM role is built on earned trust and respect. We knew from the beginning that modeling this trust and respect for CoP members was key to our success. Further, we focused on establishing trust within the CoP and continually reminding members that what is said in the meeting stays in the meeting. This does not mean that things are discussed that are not appropriate, but that there is an opportunity to discuss and clarify an issue with colleagues to determine solutions and approaches without pointing fingers or causing alienation of others. The importance of this aspect to ultimate success—trust and respect—cannot be overstressed.

Following is an example of how our use of CoP strengthened and created networks that led to increased knowledge sharing and produced outcomes that enhanced efficiency and effectiveness, as we hoped. But these CoP went even further, and when the agency went through the third, very painful and extremely disruptive downsizing in the 2008 to 2010 time frame, the value of the CoP and the networks they created became very obvious in their ability to keep focused and more quickly develop a plan for how to manage given significant changes in management needs while quickly on-boarding employees to new positions and responsibilities. It should be noted that this third set of cuts began in the engineering and inspection ranks, leading to loss of a large component of learned experience in a DOT environment, just before the workload sharply increased.

We did not start out with this collection of CoP in mind, rather they grew organically. In this case, the four groups that evolved came together over time and developed strong knowledge-sharing practices that crossed geographic and organizational boundaries. Each CoP has several things in common:

- Statewide representation
- Management support
- Grassroots participation
- Focused agendas directed toward outcomes

Construction Quality Managers Communities of Practice (CoP)

Construction inspectors are the boots-on-the-ground first line of defense in the state's ability to ensure that the work done to build or maintain the state's roadways is done safely and that we produce a product that is a good investment of taxpayer dollars. These are the people on the construction site who are trained to make sure that the right materials are being used in the right way and at the right time. These are the people who can make the difference between a road safe to drive on at highway speeds in the rain—or not, and the difference between a road that breaks down in 3 years instead of 10. Construction inspectors have to know the policies, procedures, and requirements that address their area of expertise and to apply that knowledge to construction in real time in the field. This expertise does not come easily or quickly, and only comes in relatively unique settings. A Virginia highway inspector has to know conditions that may be different than those in New York, and that are different, at least in some respects, than those in vertical construction (buildings). It takes years of training and experience to make a senior inspector. In previous times, this was the entry-level position into the agency as it provided new employees with a very hands-on and broad experience.

KM initiated this CoP (Figure 8.2) on hearing from multiple sources of the critical importance of the quality inspection program. The CoP began with a representative at the senior inspector or construction manager level from each of the nine districts and with a representative of the central office construction quality program. Over time we had slightly more or fewer people, but the size stayed at less than 15. They met quarterly to discuss and share experiences about what was happening across the state and to identify opportunities within their control that they could initiate and that would support inspectors statewide in assuring the quality and safety on VDOT projects.

The CoP identified its purpose as "appropriate improved quality efforts across the Districts for the purpose of ongoing and continuous improvement in day-to-day operations." They explicitly chose to "start with achievable near term goals and build a foundation of incremental successes." The group had several accomplishments, including the following two examples:

- Developed the concept of a single interface for the manuals and guides needed on the job and that included a consistent filing system. This initiative won a Commissioner's Award.

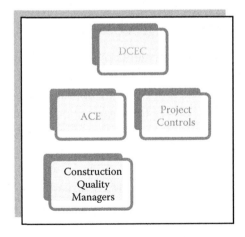

FIGURE 8.2
Construction quality managers community of practice (CoP).

- Developed, implemented, and maintained a statewide lessons learned program (now in its fourth year) that gets consistently high marks from the primary audience.

Area Construction Engineers (ACE) CoP

As a result of an audit calling for a larger presence in the field of licensed professional engineers (PEs), the role of the area construction engineer (ACE) was implemented in 2005 and 50 PEs were brought in to the agency. ACEs were given the responsibility to oversee engineering decisions that needed to be made in real time, in the field. In addition to knowing the explicit knowledge needed, the ACE must be prepared to exercise professional judgment about the right solution at the right time. The results of this professional judgment can have not just quality but monetary impacts. This CoP (see Figure 8.3) was formed when a group of ACEs, hearing about the construction quality managers CoP, came to KM and asked for help in forming a CoP to allow for collaboration and to create a statewide presence for the ACEs.

When the ACE CoP was formed, the agency had ACEs for 2 years, but the roles and responsibilities were still not clearly understood or settled.

The objective and purpose statement for forming this group provides a good statement of the role of the group, a statement that continues to be used as a framework for meetings and activities of the group. This group plays a pivotal role in accomplishing construction in the agency, and this

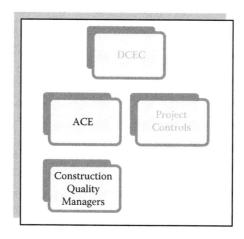

FIGURE 8.3
Area construction engineers community of practice.

CoP is meant to support that role through the sharing of information to enhance consistency and efficiency of operations. As appropriate special projects suggest themselves through the sharing of information, the group will decide whether the project is an activity that this group wants to pursue.

In most cases when we form a CoP we want to keep it small, in the range of 10 to 15 people to encourage open and frank discussion. In this case, because the ACEs were such a new group, we invited all ACEs statewide to participate as members. In addition, we have a representative from the central office division who provides construction-related policy.

The CoP provided a means of defining the ACE role, of developing a statewide voice, and of identifying a key group for review of agency-wide changes to policies that impact construction. It is a given now in the agency that this is the group to come to for feedback about field impacts of proposed policy changes.

Project Control CoP

The success of the construction quality managers and ACE CoP in focusing the efforts of these two groups led to a request by VDOT construction managers to create a project control CoP (see Figure 8.4). The purpose of the project control CoP is to serve the department in achieving its core business metrics with a focus upon delivering the construction and maintenance programs on time and on budget.

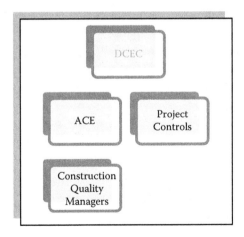

FIGURE 8.4
Project control community of practice.

The project control CoP was formed specifically to look at managing the information needed to meet contractual requirements. For example, an early project involved developing a project tracking tool that provided in one place and at the right level of detail the project data needed by project managers on a regular basis to monitor projects.

District Construction Engineer (DCE) CoP

The district construction engineer (DCE) CoP, normally called the DCE Council or DCEC, is made up of the person in each of the nine districts who is responsible for the construction program in that district, and the state construction engineer, for a total of 10 people in the state with the primary responsibility for VDOTs construction program. In 2008, their joint responsibility was equivalent to about $3 billion.

The people who initially formed this group brought significant depth and breadth of experience, both in highway construction and at VDOT. Prior to forming the CoP, the group had been meeting for about a year in a loosely organized fashion to discuss issues that impacted the statewide program. They met quarterly in person and used video-teleconferences (VTCs) as needed between the in-person meetings. Because these were the people who were primarily responsible for VDOT's construction program it quickly became known as the place to go to if you wanted to push information to construction. The meetings started growing so that within a year they turned into very large meetings to which people invited themselves. The 10 primary participants would be at the table with as many as 30 to 40 people sitting in a ring around them. Most of the time ended up being taken by people presenting information that could more effectively be provided via e-mail with very little or no time left for discussion among the 10 principals.

The DCEs decided they needed more formal organization and more focused (and smaller) meetings in order to be effective as a statewide group. All of the members of the CoP noted above reported to the members of the DCEC, either directly or through other organizational layers. Having seen the successes coming out of the CoP (see Figure 8.5), the DCEC asked KM for support in accomplishing this change. KM facilitated a strategic planning process that enabled the group in identifying a mission, objectives, and initiatives that were firmly embedded in the agency's business objectives and that were tactically embodied through a focused approach that would most support their ability to deliver the construction program.

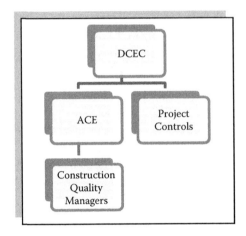

FIGURE 8.5
District construction engineers community of practice.

As a result of this strategic planning process, the DCEC identified changes in three areas: meeting organization, focused initiatives, and more closely integrating the CoP in the construction arena.

The DCEC used selected initiatives as a framework to more closely integrate all of the construction-related CoP. Each CoP continued to be expected

- To take time to discuss the issues they were facing in their areas of responsibility
- To support each other in finding solutions in their area of control
- To share best practices
- To bring issues to the DCEC that they felt important but that were outside of their area of control
- To develop recommendations for issues assigned to them by the DCEC

A DCE representative was appointed for each of the CoP, not to attend every meeting, but to be a conduit for questions and answers. Finally, the first half hour of each DCEC meeting is taken up with a VTC in which the chairs of the CoP provide a status on their activities. This monthly half-hour means that not only the DCEC, but also all of the CoP chairs should find out what is going on and can identify opportunities to leverage activities across the groups.

One of the most important outcomes of this closer connection was to bring a level of middle and upper-middle management support to their overall efforts. This focus resulted in a clear understanding within the agency that the construction arena has been able to effectively cross silos and collaborate. This group is pointed to as an example of the districts and CO working together.

Accomplishments

Because of the focused statewide perspective of the construction CoP, they were successful in providing a clear locus for construction issues which includes both the Central Office (CO) and the districts:

- They provided a clear avenue for CoP at lower levels in the organization for questions and answers while expanding networks of people in the construction line of business for the agency.
- They built a common store of knowledge and continue to do so. Two examples include a statewide filing system and an agency-wide lessons learned program that won a national award based on Baldridge Quality Criteria. This common store of knowledge was able to survive the loss of individuals and their expertise.*
- They provided a single statewide accumulation of expertise capable of responding efficiently to problems and to providing solutions and insight. While this kind of response did occur before, this set of CoP made that response much more efficient and effective. People know where to go to get a consensus statewide perspective.
- Through the collection of CoP, the construction function was extremely effective at breaking holes in silos and in supporting each other in delivering the program.
- The CoP also proved to be a great opportunity for staff to develop collaboration skills and an increased understanding of statewide needs that transcend those of their immediate work area.
- Because of their existing organization and practices, they were able to respond more quickly to changes than other areas in the agency.

This last bullet became critical after the third major, painful, and heavily disruptive downsizing that occurred at the agency in the 2008 to 2010 time frame. The agency lost a significant number of staff, mostly newer

---

* For more discussion about organizational memory, see Hammer, M., and Clark, K. (2009). Organizational memory. In *Encyclopedia of Library and Information Sciences* (3rd ed.). New York: Taylor and Francis.

and younger staff. In an agency with a 100-year history like VDOT, newer could mean someone with less than 10 years in service. In addition, because engineering was one of the first areas to be downsized, just before new funding led to a sharply increased workload, there was a sharply increased outsourcing of work that occurred. As noted previously, transportation construction work is a relatively unique niche with its own requirements and constraints, much of which can only be developed on the job and over years. In addition, churn in the agency as people left, first from the downsizing, then because of offers from the private sector as outsourcing increased, and later because of moving to some of the newly opened positions meant that there were a lot of people in new positions in which they had very little experience. For example, the DCE group had seven of the nine people replaced within 1 year.

However, and this is one of the biggest accomplishments of the CoP framework, because of the strong and effective communities of practice already in place, new people were able to quickly jump in to existing networks and to learn from their colleagues—and avoid that loss of connections between knowledge blocks experienced in 1991.

This was particularly important as the agency changed, virtually within 3 years and with very little planning, from one that essentially oversaw and completed work with its own staff to one that had to manage consultants in accomplishing that work. Such a change means doing work differently and means new skill sets. Under no circumstances would this kind of change be seamless. However, because of the organization in place, the DCEC and its CoP already began to focus and identify changes that need to occur to respond to this change, rather than continuing to try to manage in changed circumstances using the old methods.

LESSONS LEARNED

As noted earlier, an important element to our success was to have KM as a neutral facilitator trusted by the CoP to keep the group focused on the big picture and business objectives that they can directly impact.

Early successes are important to participation by CoP members. People may not know what they know or how to share what they know or with whom. In addition, sharing may be seen as too difficult or time consuming; this is particularly prevalent in government organizations faced by downsizing in financial and human resources. As simple as this may sound, our early work was really focused on these fundamentals. In the first year CoPs were formed at our instigation. Early successes resulting from CoPs reinforced their value so much that after the first year all of our CoPs were formed at the request of customers in the agency who had seen the results from CoPs.

Early successes are also important to support by management. Backing by the top executive team was vital, particularly in the early days of the program to ensure that employees were given the support to participate. But it was necessary and important to actually demonstrate the value of building grassroots support within the agency through these CoP to continue support by management.

Success in the CoP is tightly tied to keeping a focus on business objectives and on outcomes. We continually hold the groups to the notion that they are not just meeting to share their experience. Instead, they need to keep focused on identifying and responding to issues that arise and that impede their ability to do their job. This certainly includes sharing information, but it has to go beyond sharing information to using that information. If members of the CoP are not getting information that they can take back to their districts and that support them and their colleagues in doing their job, then it is not a CoP and we do not continue it.

CONCLUSION AND FUTURE RESEARCH NEEDED

According to Gupta et al., (2000), "organizational culture itself prevents people from sharing and disseminating their know-how in an effort to hold onto their individual powerbase and viability" (Knowledge management section, ¶3). This is especially true in a culture that experienced traumatic change and a loss in trust and stability. Establishing communities of practice that provided strong and quick wins with recognition for members and an immediate return on the time they invested helped to counter that loss of trust and to demonstrate the value in collaboration and knowledge sharing.

Communities of practice played a clear and significant role in creation of a knowledge-sharing culture and in network formation. The integration of the CoPs in the construction arena has shown their value in increased effectiveness and efficiency on an ongoing basis. It also showed its value in preparing the construction function to deal with very trying disruptive change.

Waiting until the disruption in your networks has occurred to develop a mechanism to keep them intact means having to play catch-up. Even with the thoughtful process taken during the 1991 workforce reduction, the networks were negatively impacted. Previously, the networks were a by-product of the apprenticeship approach so as that apprenticeship was disrupted, the networks did not develop completely and knowledge sharing was negatively impacted. With time, the impact to the networks solidified into silos.

Contrast this with the ability of the construction communities of practice that were in place during the last major downsizing, to be ready to identify and to begin to address needs. This last phenomenon suggested to us that the best preparation for dealing with major disruptive change was to have a knowledge-sharing culture in place, and that CoP, when used rigorously and held to focus on outcomes, can promote such a culture.

Future research may include how to integrate networks across lines of business within an organization, in our case across project development, construction, maintenance, and operations, and across partner organizations. It would also be of interest to more fully research successful onboarding of new employees into an organization using these networks to determine what additional value could be realized as the new more mobile generation enters the workforce.

REFERENCES

Argote, L., and Ingram, P. (2000). Knowledge transfer: A basis for competitive advantage in firms. *Organizational Behavior and Human Decision Processes*, 82, 150–169.

Augier, M., and Vendele, M.T. (1999). Networks, cognition, and management of tacit knowledge. *Journal of Knowledge Management*, 3(4), 252–261.

Burk, M. (2000, May/June). Communities of practice. *Public Roads*, 18–21.

Connelly, C.E., and Kelloway, E.K. (2003). Predictors of employees' perceptions of knowledge sharing cultures. *Leadership and Organization Development*, 24, 294–301.

Cross, R., Davenport, T.H., and Cantrell, S. (2003, Fall). The social side of performance. *MIT Sloan Management Review*, 20–21.

Davern, M. (1997). Social networks and economic sociology: A proposed research agenda for a more complete social science. *American Journal of Economics and Sociology*, 56, 287–302.

Goffee, R., and Jones, G. (1996, November–December). What holds the modern company together? *Harvard Business Review*, 133–148.

Granovetter, M.S. (1973). The strength of weak ties. *American Journal of Sociology*, 78, 1360–1380.

Gupta, B., Iyer, L. & Aronson, J.E. (2000). Knowledge management: Practice and challenges. *Industrial Management & Data Systems*, 100, 17–21.

Lesser, E., and Prusak, L. (2001). Preserving knowledge in an uncertain world [Electronic version]. *MIT Sloan Management Review*, 43, 101–102.

Okhuysen, G.A., and Eisenhardt, K.M. (2002). Integrating knowledge in groups: How formal interventions enable flexibility. *Organization Science*, 13, 370–386.

Smith, H.A. & McKeen, J.D. (2002, April). Instilling a knowledge-sharing culture. Paper presented at the meeting of the Third European Conference on Organizational Knowledge, Learning, and Capabilities, Athens, Greece.

9

Social Network Analysis: A Pharmaceutical Sales and Marketing Application

Molly Jackson, Doug Wise, and Myra Norton

INTRODUCTION

How do we sell more? How do we acquire more customers? How do we build brand loyalty? Sales and marketing departments at companies across every sector ask these questions every day. Segmentation analyses are a primary tool to help answer these questions. By determining the different behavioral and emotional motivations of a particular customer base, sales and marketing teams identify unique strategies to prioritize and target consumers. Some segmentations group target consumers by their tendency to adopt a given type of product (e.g., "innovators," "early adopters," or "laggards"),[1] while others identify their brand loyalty ("entrenched," "convertible," or "available").[2] However, these strategies only lend themselves to direct sales and marketing techniques. In addition, behavioral and emotional characteristics give insight into only one dimension of a consumer's decision-making process. The critical question remains: Why or how does someone come to believe something about a certain brand or market? The answer to this question is the key to understanding a customer's purchasing decisions, and ultimately will allow companies to engage with customers in an authentic manner, improve their product and service offerings, and thereby meaningfully influence customers to choose their brand.

More and more each day in every market you can think of, customers are overloaded with information from many different channels. These channels are continually growing and changing with the recent wave of novel social marketing technologies. When making a purchasing decision, this accumulation of information can be overwhelming. Consumers very often turn to people they trust to help them distill, synthesize, and interpret the information, thus reducing the risk and uncertainty associated with their decision. In turn, the peers these customers turn to for advice also seek guidance from other people whose expertise they trust. Advice has a fascinating way of streaming from person to person, branching across a complex network of personal and professional ties. These systems of advice are called *social networks*, and they are the most powerful force in the decision-making process (Figure 9.1).

The study of social networks has academic foundations beginning with Everett Rogers' study of the diffusion of innovations. He defined diffusion as "The process in which an innovation is communicated through certain channels over time among the members of a social system."[1] There is naturally some degree of associated uncertainty and perceived risk with any novel idea (product/service); however, an individual can reduce this degree of uncertainty by obtaining information.[1] How and from whom individuals obtain this information is the crux of social network analysis.

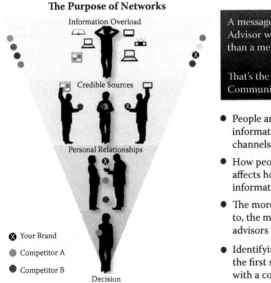

The Purpose of Networks

Information Overload

Credible Sources

Personal Relationships

⊗ Your Brand
● Competitor A
● Competitor B

Decision

A message delivered from a Trusted Advisor will always be more powerful than a message from a biased source.

That's the basic idea behind Community Analytic's approach.

- People are bombarded with information from a multitude of channels
- How people are connected socially affects how they access and interpret information
- The more information they are exposed to, the more they rely on trusted advisors to distill and interpret it
- Identifying these trusted advisors is the first step in successfully engaging with a community

FIGURE 9.1
The purpose of social networks.

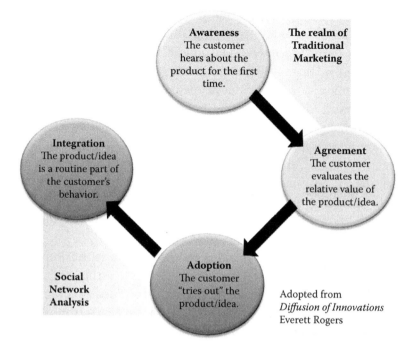

FIGURE 9.2
Moving from traditional marketing to social network analysis.

Although traditional marketing focuses on customer awareness and agreement with a product's value proposition, these alone are often insufficient to move someone to adoption and integration of the product into his or her routine behavior. Social network analysis focuses on revealing the authentic, preexisting channels that individuals use to obtain information, vet ideas, and assure relevance for their needs, which in turn reduces their feeling of risk and leads to adoption and integration into their routine behavior. These channels can be leveraged to more effectively spread ideas and spur community action (Figure 9.2).[1]

BACKGROUND ON COMMUNITY ANALYTICS®

Community Analytics (www.CommunityAnalytics.com) was founded upon the importance of social networks. It is the second generation of a firm, COMSORT, that mapped influence networks among physicians for many pharmaceutical companies in the 1990s. In 2001 COMSORT entered into an exclusive partnership with Merck, and was acquired by Merck in 2003. The founder of COMSORT and Community Analytics, John Hawks, built the operational model for implementing this type of data globally. During the tenure of the company's noncompete with Merck, Community Analytics developed new methodologies and analytics while working in other industries, such as financial, legal, information technology, and higher education. The company reentered the healthcare industry in 2008 and is innovating and refining data collection methodologies, analytics, and implementation for pharmaceutical, biotechnology, healthcare systems, and other companies in healthcare. Even though the company continues to work in a variety of industries, we will use the healthcare industry as the context for this exploration of social networks.

One of the powerful innovations in the pharmaceutical space is a methodology developed by Community Analytics that engages and involves a company's pharmaceutical sales force, medical affairs team, and other internal resources throughout the process—from data collection to implementation. Community Analytic's current healthcare clients are experiencing great success with this new approach. The involvement of company representatives in the data collection component results in deeper relationships between the company and their customers and provides an opening for many representatives to gain "face time" with hard-to-see physicians. This process generates buy-in and alignment among sales, communications, marketing, and medical affairs departments from the beginning stages of client engagement. In addition, Community Analytics supports sales, communications, marketing, and medical affairs throughout the process to assure that every person involved has a solid understanding of the outcomes we are trying to achieve and the role they play in helping to achieve those outcomes.

Let's look at a real-world example.

CASE STUDY

PharmaCompany, which develops pharmaceutical products in the acute care environment, came to Community Analytics in need of a new sales approach for one of its products. Upon its launch in 2003, Product X experienced significant growth and maintained impressive and consistent sales into 2008. However, Product X was facing a widely used competitor that was much less expensive and had already been adopted on most formularies. While Product X was deemed a superior treatment, PharmaCompany's challenge was convincing physicians to change their current course of treatment for one that was more expensive. For the most part, PharmaCompany used very traditional sales techniques, including in-office detailing of the product's efficacy, safety, tolerability, mechanism of action, and so forth, providing drug samples, and offering other scientific medical information. Physicians were targeted through traditional means, such as segmentation analyses and prescribing behavior. With an impending patent expiration and persistent competition from its less-expensive competitor, PharmaCompany was interested in developing a new approach to educating physicians and, ultimately, increasing Product X prescribing.

SOCIAL NETWORKING PRINCIPLES APPLIED

The approach outlined uses Community Analytics' systematic methodology to connect brands with existing communities. This four-step strategic framework, specifically applied to PharmaCompany's business goals of increasing Product X prescribing, is outlined in Figure 9.3.

STEP 1: OPPORTUNITY IDENTIFICATION

Serving as the foundation of this new approach to boosting Product X sales, Step 1 involves mapping the networks of trust and advice seeking among the target audience (in this case, physicians) and identifying the Key Network Members™ (KNMs™), those people who hold strong strategic positions within the network. These KNMs may possess the ability to reach large portions of the network, bridge disparate groups, and serve as trusted advisors to many network members.

Exercise 1: Map the Network

Traditionally, Community Analytics, and other similar social network analysis firms, map networks of physicians (as well as other types of targets) by conducting primary research among the target audience, often through online, mail, or telephone surveys. In this case, Community Analytics utilized its unique approach of using sales representatives and medical affairs representatives to gather the data directly from physicians during their routine visits.

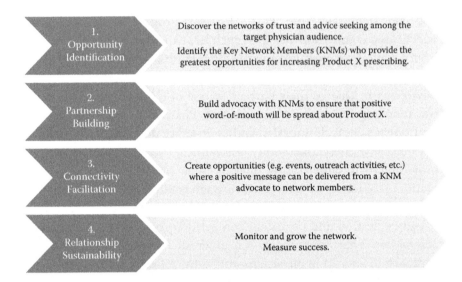

FIGURE 9.3
Community Analytic's strategic framework for PharmaCompany.

This methodology holds many benefits to traditional research. It results in more substantive relationships between the company and physicians, provides an opportunity for sales representatives to gain "face time" with hard-to-see physicians, and generates internal company buy-in from key stakeholders (including sales, communications, marketing, medical affairs, etc.) from the beginning of the engagement. In addition, while physicians are routinely paid an honorarium in return for their participation in market research surveys, including social network surveys, this methodology allowed PharmaCompany to appreciate a substantial reduction in research costs due to the elimination of these honorarium expenses. More importantly, the incentive to the physicians to participate in the research was an intrinsic one—the desire to improve educational events as well as interactions with sales and medical affairs representatives. Finally, as an added benefit, the overall response rate using Community Analytics' unique data collection methodology was significantly higher than is traditionally seen in more conventional physician network surveys (30.2% overall response rate versus 13.4%[3]), which is especially significant for network studies in particular, where the value experienced from mapping a network increases as the response rate increases.

In this case, to map the network, physicians were asked by their PharmaCompany sales and medical affairs representatives to complete a needs assessment. This assessment asked them to name up to three topics/issues of interest related to infectious disease, as well as the people they feel are best suited to moderate a discussion on these topics (their "nomination"). In addition, physicians were asked to indicate their (trusted advisor for a prefilled topic of specific interest to Pharma Company.) Data were collected in 18 sales regions across the United States.

Exercise 2: Identify Key Network Members (KNMs)

Community Analytics uses a sophisticated program of direct communication, technology, and mathematics to accurately map the connections that matter most to a community, from the community's perspective. As this research progresses, a complex and relevant influence network evolves. In cases where company representatives are trained to gather network data, Community Analytics employs a quality control mechanism to validate responses and eliminate any associated bias. In addition, responses are cleansed and de-duped prior to applying network analysis (Figure 9.4).

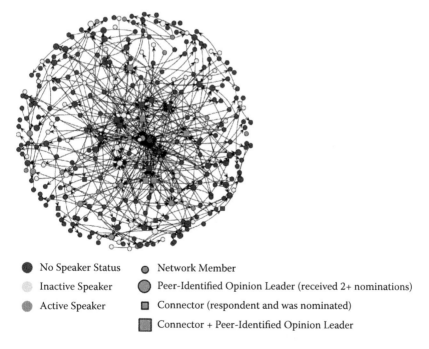

● No Speaker Status   ○ Network Member

● Inactive Speaker   ◐ Peer-Identified Opinion Leader (received 2+ nominations)

● Active Speaker   ▢ Connector (respondent and was nominated)

▪ Connector + Peer-Identified Opinion Leader

FIGURE 9.4

Example PharmaCompany network map for Region 1 across all topics.

After a network is mapped, social network analysis is conducted to identify and prioritize network members who hold strategic positions in the network. We call these individuals Key Network Members (KNMs). Among the KNMs, those individuals who received the greatest number of nominations are identified as Peer Identified Opinion Leaders™ (PIOLs™). These individuals are essential to the process because they speak to and speak for the community as a whole, and are the conduits through which an organization can establish a dialogue with a community. We also identify Connectors™, individuals who grow the network as a result of their links between members and their reputation as a trusted source of information. Finally, any network members who serve an advisory role to at least one other network member are considered Trusted Advisors™.

When reviewing the network data for Product X, the following key questions were asked:

1. Who are the Key Network Members?
   a. Who are the Peer Identified Opinion Leaders?
   b. Who are the Connectors?
2. Who are the Trusted Advisors for high-priority physician targets and accounts?
3. What are the most named topics of interest among physicians?

Analysis revealed that, across all topics, a total of 976 physicians were nominated as potential speakers (i.e., Trusted Advisors), with 423 physicians being PIOLs (receiving nominations from two or more physicians). The research focused on both on-label and off-label topics of interest. Across on-label topics, a total of 567 physicians were nominated as potential speakers, with 231 physicians receiving on-label nominations from two or more physicians (Figure 9.5).

Exercise 3: Conduct a Gap Analysis

Once the data are mapped, gap analyses are often conducted to assess the company's current awareness of Key Network Members. This analysis reveals weaknesses in a company's current targeting strategies by identifying the KNMs that a company is currently not targeting or focusing its efforts on, and the physicians a company is targeting but who are not as influential according to the network members.

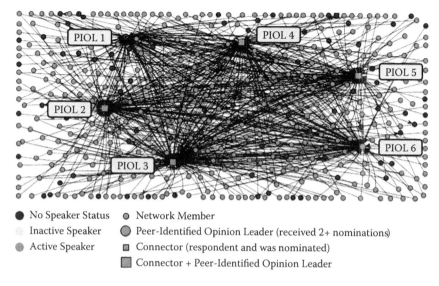

● No Speaker Status   ◎ Network Member

◌ Inactive Speaker   ◉ Peer-Identified Opinion Leader (received 2+ nominations)

◍ Active Speaker   ▫ Connector (respondent and was nominated)

                ▪ Connector + Peer-Identified Opinion Leader

FIGURE 9.5 (See color insert.)
Example network map: national network map of the top Peer-Identified Opinion Leaders for on-label topics.

For PharmaCompany, the network data were matched against several of their existing databases to determine each KNM's status. The following specific analyses were conducted at both national and regional levels:

1. Medical Affairs Key Opinion Leaders (KOLs) gap analysis
2. Active Speakers gap analysis
3. Current physician targets gap analysis

For illustration purposes, let's look closer at the gap analysis conducted among PharmaCompany's current physician targets. At the regional level, a gap analysis was conducted on all physicians who received two or more nominations from the given region, regardless of the physician's location. The networks were analyzed across all topics and against all on-label topics. Analysis revealed key failures in PharmaCompany's current physician targeting: one-third of the PIOLs identified were not currently being targeted by Product X representatives (Figure 9.6).

This illustrates an opportunity for PharmaCompany to reprioritize their targeting strategies, focusing on more influential physicians, as a message delivered from a trusted and unbiased source is *10 times more powerful* than traditional marketing messages (Figure 9.7):

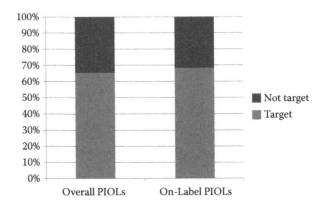

FIGURE 9.6
Current physician targets gap analysis.

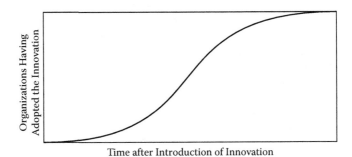

FIGURE 9.7
Bass curve adoption equation.

where N_t is the number of adopters, m is market potential, p is the coefficient of external influence (media, sales force, etc.), average value equal to .03, and q is the coefficient of internal influence (word-of-mouth, friends, relatives), average value equal to .38.[3]

Exercise 4: Conduct Individualized KNM Assessments

This exercise entails taking a closer look at each individual Key Network Member. Here, each KNM's unique network map and personal attributes are reviewed to assess

- Whom do they respect and who respects them?
- What topics are most important to them?
- What are their attitudes/beliefs toward Product X: positive, neutral, or negative?

- What is the appropriate reach and frequency of sales calls?
- How should the sales manager be involved in connecting with this KNM?
- Should other senior management be involved?
- How should Medical Affairs be involved?
- Is the KNM influencing more than one area or region?
- Is the appropriate level of cross-territory/region/area coordination occurring?

In addition, assessment of the KNMs will reveal areas where resources need to be reallocated. The following key questions can help in this process:

- Are the influencers for high-priority physicians being targeted?
- Are the most respected speakers utilized the most; if not, why not?
- Are speakers facilitating the right topics?
- Are the right physicians on the invite list for Promotional Medical Education (PME) programs?
- Are there KNMs who should be added or removed from the Medical Affairs' KOL list?

After these four key exercises are completed, the areas of opportunity are identified, and now *Partnership Building* begins.

STEP 2: PARTNERSHIP BUILDING

The goal of this second step is to build advocacy with Key Network Members to ensure that positive word-of-mouth will be spread about a product. This step does not require dramatic changes to current activities. Rather, the addition of influence network data simply informs and improves current initiatives while simultaneously inspiring new approaches. The power of building partnerships is that it enables organizations to communicate effectively with and listen to the community. It is important to note that the path to building a partnership may be unique for each KNM. Understanding and building partnerships unique to each KNM based on their needs and desires is paramount.

In the PharmaCompany example, Community Analytics worked directly with PharmaCompany during multiple strategy sessions to brainstorm tactics for building positive relationships with KNMs and to create action plans.

Exercise 5: Reprioritize Targets

In order to build partnerships with key targets, it is first important to prioritize the KNMs. The following factors can help in this prioritization:

1. Strategic value to the network (prioritize individuals who influence the most network members, and those who are Connectors, connecting disparate network groups)
2. Level of advocacy for the company's product

PharmaCompany reprioritized its sales strategy in several ways. First, it developed strategies for targeting the Trusted Advisors of their key accounts. In addition, it improved their speakers' list by prioritizing individuals the network members deemed most capable of educating them on particular on-label topics of interest. Finally, each salesperson augmented his or her target list by focusing on those physicians who add strategic value to the network in their territory: those who are PIOLs and Connectors (e.g., physicians influencing multiple hospitals or physicians in their territory).

Sales representatives also ranked their perception of each KNM's advocacy for Product X using a Likert scale (1 = Product X Negative to 5 = Product X Advocate). Then, representatives tailored their partnership building to move KNMs toward greater levels of advocacy.

1. Low Advocates (1 to 3 rating) were targeted with tactics to improve awareness of, and agreement with, the advantages of using Product X and to increase trial of Product X.
2. Focus for Medium Advocates (4 rating) was on maintaining their personal positive experiences and reinforcing positive outcomes they witness among patients.
3. High Advocates (5 rating) were targeted for the next step in the process, Connectivity Facilitation, in which they are asked to support PharmaCompany through programs that allow them to share their positive Product X experiences with their peers.

As with any market research initiative conducted by a sales force, it is very important that the salesperson share the results with the participants. There are two reasons for this:

1. It shows the respondent that the company is listening to him or her and taking action on what the respondent said.

2. It allows the salesperson to have a value-adding conversation with the respondent about the results.

In this example, a brief flyer listing the top locally relevant topics mentioned by respondents was distributed and discussed with all physicians in each territory. It is important to note that even physicians who did not participate in the research were given a copy of the results. This builds trust with nonparticipating physicians and encourages them to participate in future research initiatives.

STEP 3: CONNECTIVITY FACILITATION

Once two-way communication is established between the organization and the network, Community Analytics helps organizations connect KNMs/PIOLs to each other and to community members with common interests. The primary goal of this step is to create opportunities where a positive message can be delivered from a KNM advocate to his or her network members. Ultimately, one hopes to create an environment where the roles of "customer" and "vendor" are no longer relevant and where dialogue is focused on the community's needs, the best ideas, and the most effective plans for realizing these ideas.

Exercise 6: Connect the KNM Advocates to Their Network Members

The goal here is to connect High Advocates with Low/Medium Advocates in the network. These tactics can be anything, from small, intimate events to more formal, large-scale events.

PharmaCompany initiated the following programs:

1. *Promotional Medical Education Events were organized.* Speakers were identified based on network nominations on certain relevant topics, as well as their level of advocacy for Product X (focus was on those PIOLs and Connectors who have positive opinions toward Product X). The invite list consisted of the speakers' networks, regardless of geography. If a PIOL/Connector was not a good fit as a speaker, that person was invited to host a table at the event with his or her network. This ensured attendees were able to interact with their trusted peers one-on-one.

2. *Round Table Discussions covering the most prevalent topics were planned at industry events.* Each table at the event was purposely seated based on the networks uncovered in the network research. This allowed physicians to sit with their trusted peers and discuss each topic with those they value most in their network.

3. *Sales representatives kept physicians up-to-date on new publications by their trusted advisors.*

4. *Sales representatives arranged meetings between physicians and their trusted advisors.* Face-to-face, personal meetings were arranged by the sales representatives in which they brought along PIOLs (who were advocates of Product X) to personally meet physicians who trust and want to engage with them.

STEP 4: RELATIONSHIP SUSTAINABILITY

The Community Analytics team provides ongoing support to monitor and grow the network. This is fundamentally different from traditional marketing approaches that focus on executing and evaluating "campaigns." Here, time is taken to evaluate and measure the success of activities, and adapt new strategies to continually engage the network over time.

This fourth step in the strategic framework can include the following exercises:

1. Continue to identify new opportunities.
   a. Monitor and record new relationships that are uncovered (perhaps through sales visits).
   b. Augment the network map.
   c. Continually review strategies based on the refreshed network map.
2. Track KNM advocacy as it shifts over time using the same Likert scale and adapt outreach as necessary.
3. Evaluate outreach activities.
   a. Event satisfaction (compare events utilizing social network strategy versus not)
   b. Event attendance (compare events utilizing social network strategy versus not)
   c. Postevent prescribing behavior (compare events utilizing social network strategy versus not)

    d. Current measurements (e.g., account/territory/region/area prescribing numbers)—compare before/after social network strategy in place

4. Maintain, build, and track partnership activities.

    a. Identify goals for closure of gaps (Targets, KOLs, and Speakers) and measure success in meeting goals.

5. Track Product prescribing to assess the value of this approach.

6. Track implementation of the network approach.

    a. Track long-term social network action plans for KNMs.

    b. Gather ongoing feedback from the physicians on the progress of taking action on the survey results (ideally via KNMs).

    c. Create quarterly advocacy goals: metrics/measurement of advocacy.

    d. Monitor targets for top KNMs in region—keeping thumb on the pulse of top KNMs.

FUTURE RESEARCH NEEDED

Networks evolve slowly. We typically see a 5% change in influence networks per annum. An influential individual remains influential barring any untoward behavior. New product entrants to the market and additions of new network members are other examples of factors that may transform a network. These are more likely to occur with greater frequency. Therefore, it is important to continually monitor and record new relationships and augment the network map. This continual analysis of the network will allow a company to continually evolve its targeting techniques in the most strategic and effective ways possible. We recommend an annual update to network data for most organizations.

LESSONS LEARNED

Mapping influence networks and applying them to your business is an exciting endeavor. However, with this task come unique challenges. The following key takeaways should be remembered when beginning any networking mapping project, and more specifically one in the health-care arena:

1. Understanding the "why" behind a customer's perceptions of a brand is the key to understanding the customer's purchasing decisions, and ultimately this will allow you to engage with your customers in an authentic manner, improve your product offerings, and therefore meaningfully influence customers to choose your brand.

2. Applying an influence network approach to sales allows your company to approach targets in a new and meaningful way. As a message delivered from a trusted advisor will always be more powerful than a message from a biased source[4] (like one directly from your company), this approach allows you to determine who is influencing your potential customers and then share meaningful messages with those key people, allowing you to improve the efficiency of your interactions with your market.

3. Mapping the network is only the first step in applying an influence network approach to your business. You must be prepared to build advocacy with KNMs, facilitate connections between KNMs and other network members, monitor and grow the network, as well as measure the success of your efforts. If this is your first time using an influence network approach, you must be willing to commit to going outside your comfort zone of traditional marketing and be open to new outreach techniques.

4. In the case of the pharmaceutical sales industry specifically, educating and involving sales and medical affairs representatives is critical to the success of implementing this type of intelligence—the people who will be acting on the data should be involved from the beginning. You can have the best data in the world and if people do not do anything with it, it is worthless, so their buy-in and involvement are vital from the start. This methodology also improves response rates, thereby enhancing the value of your network mapping.

5. As with all research, understanding how the data will be implemented on the back end should inform how data are gathered and analyzed on the front end. Along these lines, it is important to remember that networks have context. In the case of Product X, several networks were mapped with respect to physicians' trusted advisors for several different therapeutic categories. So, remember, begin with the end in mind.

REFERENCES

1. Rogers, E.M. (1962). *Diffusion of Innovations*. New York: Free Press.
2. Hjelmar, Ulf. (2005). The concept of commitment as a basis for social marketing efforts: Conversion model as a case, *Social Marketing Quarterly* 11: 2.
3. Community Analytics historical database.
4. Bass, Frank M. (1969). A new product growth model for consumer durables, *Management Science* 15(5): 215–227.

10

Collaborating Using Social Networking at Price Modern

Gloria Phillips-Wren and Louise Humphreys

INTRODUCTION

Building close customer relationships, providing services that add real value to customers, and being recognized as a leader in the community are strategic goals of Price Modern LLC, a $120 million office furniture dealership that represents over 200 different furniture manufacturers, giving them the ability to work with a wide variety of vertical markets and customers. As a company with strong partners and some additional resources, Price Modern has the ability to leverage social media as a value-added differentiator geared not only toward existing and potential customers, but toward the large community of influencers involved in their projects. These influencers range from architects and interior designers who are planning offices and specifying furniture, to developers and brokers, who are increasingly providing consulting services or wrapping the furniture purchase into their leases as a way to generate additional revenue. This chapter presents social media as a key value-added differentiator opportunity for a small business-to-business organization, and the insight that technology must not only align with strategic business goals but with the corporate culture as well. We examine the introduction of social media into Price Modern initiatives in their "Best Practices" initiative, sales force development, and information flow to existing and new customers in order to maintain and build the trusted relationships relied on for decades in the new and socially expected way.

"Either business gets social or it gets left behind" (Austin et al., 2010). Technology allows people to collaborate and network with each other using rich media with visualization, sound, and even simulated worlds that utilize imagination and creativity (Power and Phillips-Wren, 2011). Mobile technologies deliver wikis, blogs, photos, and instant communication in real time. New intelligent technologies are changing the way that organizations create, gather, organize, and disseminate knowledge (Garrido et al., 2010; Vivacqua et al., 2011). As these technologies continue to permeate everyday life, businesses large and small are learning how to embrace and integrate these phenomena into their organizations. Social media technologies offer countless opportunities for communication and collaboration and can deliver business value to a company, its employees, its suppliers, and its customers (Culnan et al., 2010). At the same time, to realize that value, the technology needs to be aligned with the strategic goals of the organization (Rozwell, 2010; Rozwell et al., 2010). As we consider social media opportunities as applied to a small business-to-business organization, we show that it is critical to develop a plan that aligns not only with the strategic business goals but with the corporate culture as well. Development of a plan is just the first step in the process. Implementation and acceptance of the social media plan throughout an organization are another major step in the process. As we explore the addition of a social media plan into the Price Modern LLC business strategy, all of these considerations play a role.

This chapter first discusses the background of Price Modern, and then presents the social media principles used in the development of the company's social media strategy. We then discuss Price Modern's collaboration/knowledge sharing/social networking perspective followed by their social media strategic approach. The paper concludes with lessons learned and future research needs.

BACKGROUND OF PRICE MODERN

Founded in 1904, Price Modern is a $120 million office furniture dealership headquartered in Baltimore, Maryland, with offices in Washington, DC, and Raleigh, North Carolina. Price Modern has 160 employees and represents over 200 different furniture manufacturers, providing the ability to work with a wide variety of vertical markets and customers. Approximately half of the 35-member sales force has been in the industry or with Price Modern for over 20 years. The robust growth over those

20 years is a direct result of close customer relationships that have been developed and maintained. The business is regional in nature, and as far as commercial furniture dealerships go, Price Modern is one of the largest in North America.

The office furniture market is fairly small as shown in Table 10.1 (note that figures are not adjusted for inflation) (BIFMA, 2011), and it is highly susceptible to cyclical downturns mirroring the economic conditions of the country. As the historical figures indicate, it is also a shrinking market. Between tighter budget allocations, lowered panel heights on office furniture to induce collaboration, and the large amount of refurbished

TABLE 10.1

Historic Industry Growth of the U.S. Office Furniture Market

			Value of U.S. Office Furniture Market (Millions of U.S. Dollars)			
Year	U.S. Production	Percent (%) Change	Imports	Exports	Consumption	Percent (%) Change
2010	$8,300	5.8%	$2,153	$576	$9,877	7.0%
2009	$7,845	(29.7)%	$1,875	$490	$9,230	(29.0)%
2008	$11,160	(2.3)%	$2,510	$679	$12,991	(3.2)%
2007	$11,420	5.5%	$2,563	$565	$13,419	4.4%
2006	$10,820	7.4%	$2,531	$492	$12,859	7.9%
2005	$10,070	12.7%	$2,280	$438	$11,912	12.3%
2004	$8,935	5.1%	$2,022	$347	$10,610	5.4%
2003	$8,505	(4.3)%	$1,870	$307	$10,068	(2.5)%
2002	$8,890	(19.0)%	$1,777	$338	$10,328	(16.4)%
2001	$10,975	(17.4)%	$1,806	$430	$12,351	(17.0)%
2000	$13,285	8.5%	$2,094	$496	$14,883	9.5%
1999	$12,240	(0.9)%	$1,772	$430	$13,591	1.2%
1998	$12,350	7.8%	$1,532	$454	$13,428	9.6%
1997	$11,460	14.1%	$1,236	$443	$12,253	15.1%
1996	$10,040	6.4%	$968	$360	$10,648	7.7%
1995	$9,435	6.6%	$798	$345	$9,888	8.0%
1994	$8,850	8.5%	$677	$375	$9,152	9.7%
1993	$8,160	5.8%	$548	$364	$8,345	6.6%
1992	$7,710	6.7%	$440	$324	$7,826	6.7%
1991	$7,228		$394	$288	$7,334	

Source: Business and Institutional Furniture Manufacturers Association (BIFMA). (2011). The U.S. Office Furniture Market—Historical Growth (Updated March 17, 2011), BIFMA, Grand Rapids, MI. www.bifma.org (accessed March 26, 2011).

Note: Volume figures reflect the manufacturers' invoice value of new office furniture. Figures do not include refurbished (recycled) furniture (add an estimated 15%) or Ready-to-Assemble (RTA) office furniture (add an estimated $800 million).

and remanufactured furniture available from companies that have downsized, the competition for opportunities to sell new furniture is fierce.

Table 10.1 depicts market consumption that is defined as production, plus imports, minus exports. Production represents the total shipments, or sales, value of office furniture manufacturers located in the United States to all locations in the world. Consumption represents the value of all office furniture sold in the United States from all sources in the world including those in the United States.

In order to thrive in this market, Price Modern relies on building close customer relationships, providing services that add real value to its customers, and being recognized as a leader in the community. As a company with strong partners and some additional resources, Price Modern has the ability to leverage social media as a value-added differentiator geared not only toward existing and potential customers, but toward the large community of influencers involved in its projects, as well. The influencers range from architects and interior designers (from young to more mature) who are planning the offices and specifying the furniture, to developers and brokers, who are increasingly providing consulting services or wrapping the furniture purchase into their leases as a way to generate additional revenue.

As part of the value-added differentiator strategy, social media components are implemented in several ways at Price Modern: to communicate internally with Price Modern's healthcare office furniture initiative, to share information in a Best Practices initiative, to provide information to the sales force; and to interact with customers. With healthcare reform on the horizon and the rising awareness of healthcare-related issues such as infection control, there are numerous possibilities for providing valuable information and an open forum for discussions about types of furniture that could help with these issues. Price Modern is also undergoing a Best Practices initiative as a way to improve its processes and services. The ability to internally communicate the progress of this initiative and invite comments and input through social media is one opportunity, and social media provides a way to improve communication with external stakeholders in an area where it is difficult to keep up with the rapid pace of change and constant flow of new information. Social media provides a way to streamline the flow of information to give the sales force the ability to keep their existing customers informed while allowing them time to reach out to potential new customers in order to maintain and build the trusted relationships that have been relied on for decades in the new and socially expected way.

COLLABORATION/SOCIAL NETWORKING PRINCIPLES FOR PRICE MODERN

Social networking involves human social behaviors that are encouraged or enhanced by technology (Power and Phillips-Wren, 2011). In a business environment, these behaviors and technologies need to be related to business use that will ultimately deliver business value. Although there are many frameworks for social media, we selected the one by Bradley (2011) to develop Price Modern's social media strategy. Bradley (2011) organizes the topic around collective behaviors and relates the particular social media, business use cases, and business value to them as shown in Table 10.2.

TABLE 10.2

Collective Behaviors, Social Media Technologies, Business Use Cases, and Business Value

Collective Behaviors	Social Media Technologies	Business Use Cases	Business Value
		Brand awareness	
	Blogs	Social learning	
	Crowdsourcing	Corporate memory	
	Discussion forum	Customer service	
Collective intelligence	Idea engine	Dynamic documentation	Customer responsiveness
Expertise location	Wiki	Driving innovation	Market responsiveness
Interest cultivation	Prediction market	Event execution	Operational effectiveness
Mass coordination	Social feedback	Human relations	Product development effectiveness
Relationship leverage	Social networking	Market awareness	Regulatory responsiveness
Emergent structures	Publishing media	Operations execution	Sales effectiveness
	Virtual worlds	Product delivery	Supplier effectiveness
	Social games	Product engineering	
	Livecast	Project management	
	Content communities	Product utilization	
	Answer marketplace	Sales effectiveness	
		Tech support	

Source: Based on Bradley, A. (2011). Employing Social Media for Business Impact: Key Collective Behavior Patterns, *Gartner Research*, G00173838, pp. 1–19. With permission.

Bradley (2011, p. 5) defines collective behaviors in the following ways:

Collective Intelligence is the meaningful assembly of relatively small and incremental community contributions into a larger and coherent accumulation of knowledge.

Expertise Location is finding specific value from the masses of people and among the staggering amount of available content.

Interest Cultivation involves collecting people and content around a common interest with the goal of growing the community of interested people and increasing their level of engagement.

Mass Coordination is the rapid organization of the activities of a large number of people through fast and short mass messaging often spread virally.

Relationship Leverage is the seemingly oxymoronic practice of effectively managing and deriving value from a prodigious number of personal relationships.

Emergent Structures are structures, such as processes, content categorization, organizational networks, and hidden virtual teams, that are unknown or unplanned prior to social interactions but emerge as activity progresses.

These behaviors can be supported through *social media*, a term that includes a suite of technologies and whose definition has shifted over time (Power and Phillips-Wren, 2011). For example, Kaplan and Haenlein (2010) defined social media as "a group of Internet-based applications that build(s) [*sic*] on the ideological and technological foundations of Web 2.0, which allows the creation and exchange of user-generated content." As new technologies have been created, social media have become more fluid to allow people to interact, communicate, share, coordinate, mass together, and form new structures as suggested in Table 10.2. As such, social media is used to describe both the technology and the activity (Power and Phillips-Wren, 2011). Social media are categorized in various ways, and the categories in Table 10.2 include those by Bradley (2011), FredCavazza.net (2011), and Kaplan and Haenlein (2010, p. 61).

Table 10.2 connects collective behaviors and social media with business uses and business value. Business use cases illustrate ways that social media can be utilized within a business context. Business value can be demonstrated in terms of effectiveness and responsiveness (Bradley, 2011). The four areas can be related in a variety of ways. For example, a company with an objective of leveraging collective intelligence (a collective behavior) to increase product development effectiveness (a business value) might decide to drive innovation (a business use case) through crowdsourcing (a type of social media). Culnan, McHugh, and Zubillaga (2010) further define business value from social media in other terms such as driving traffic, viral marketing, customer loyalty and retention, cost savings, revenue, and customer satisfaction. They point out that business value is derived from the way that the technology is used, and not from the technology itself. More comprehensive studies of the business value of information technology such as that by Melville, Kraemer, and Gurbaxani (2004) include other considerations such as internal and external factors, complementary organizational resources, and the competitive and macro environment. In this paper we focus on Price Modern's use of social media for knowledge generation and sharing and connect these directly to expected business value.

PRICE MODERN'S COLLABORATION/KNOWLEDGE SHARING/SOCIAL NETWORKING PERSPECTIVE

In support of its strategic goals, the focus at Price Modern is to support the sales function to enable those teams to provide exceptional service to their customers. Consistent with past priorities, initiatives that improve and facilitate sales and customer service are valued. To that end, the focus is on management and operations, with technology viewed as a support function rather than as a strategic resource.

Price Modern's technology use has historically been on internal communication, an order entry and accounting system, and some support in proposal development and execution. Previous attempts to expand technology support in the past with applications such as Customer Relationship Management (CRM) software were not well executed, and the salesforce view is that these technologies required work with no gain, resulting in failure of the implementation. Such a result is consistent with the Technology Acceptance Model (Davis, 1989) that posits that a user's intention to use a technology has antecedents "perceived usefulness" and "perceived ease of use." Because the sales force did not perceive the system as useful or easy to use, the ultimate result was lack of use.

One way to view Price Modern's corporate technology strategy is in terms of the Strategic Grid (McFarlan et al., 1983; Pearlson and Saunders, 2010). The Grid is divided into four quadrants as shown in Figure 10.1 with the horizontal axis indicating the future strategic impact of the application,

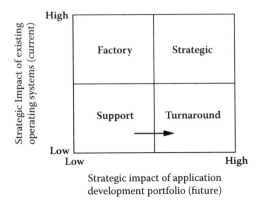

FIGURE 10.1
The Strategic Grid showing Price Modern currently in the Support quadrant of the grid and moving toward the Turnaround quadrant.

and the vertical axis indicating the current impact of the company's operating systems. The four quadrants are (1) support—low operational impact, low strategic impact; (2) factory—high operational impact, low strategic impact; (3) turnaround—low operational impact, high strategic impact; and (4) strategic—high operational impact, high strategic impact. As indicated in Figure 10.1, Price Modern is in the Support quadrant and is moving toward the Turnaround quadrant in part through the use of social media.

In order to assess how social media can move Price Modern forward to the Turnaround quadrant, we assess its strengths, weaknesses, opportunities, and threats as shown in Table 10.3. The assessment shows that Price Modern is a market leader and has the financial viability to develop a strategic social media program. The weaknesses indicate that Price Modern will need a change leader who understands both the company and the technology. Successful implementation of the social media plan is crucial because the company has had past failure with CRM technology initiatives.

Utilizing Table 10.2 and comparing the key collective behaviors to the analysis of Price Modern in Table 10.3, a strategic plan emerges for the use of social media. Bradley (2011) examined 200 cases of successful social media implementation and performed an analysis of the strongest impact patterns associated with those behaviors. Four behaviors are associated with successful social media implementations as shown in Table 10.4: collective intelligence, expertise location, interest cultivation, and relationship

TABLE 10.3

Assessment of Price Modern's Strengths, Weaknesses, Opportunities, and Threats in Considering Social Media Programs

Strengths	Weaknesses
• Largest Haworth dealership in North America • Experienced sales force and management team • Owners who support new ideas and identify need for change • Marketing team that supports the architecture and design community • Diversification in many vertical markets (government, healthcare, commercial) • Customer-focused processes and operations	• Small company means limited resources • Past successes sometimes lull Price Modern into thinking that change is not necessary • Aging sales force that does not embrace change • Aging workforce that does not embrace change • Poor internal communication that allows rumor mill and gossip to run rampant • Lack of Customer Relationship Management (CRM) tools • When unsuccessful in a competitive scenario, focus tends to be on external factors rather than looking internally to see where we can make improvements • False starts with CRM will impact the adoption of new tools
Opportunities	**Threats**
• Being a big player in a small market allows Price Modern to devote resources to developing a competitive advantage • Being a small company with very few layers allows Price Modern to be agile and respond to changing technologies • Opportunity to be early adopters of social media in this market • Financially strong and ability to invest in some advanced technologies	• Ineffective implementation of social media programs can have negative or no results • Ineffective implementation of social media initiatives will result in tools that are not used and increased doubt of workforce about programs initiated by management • Ignoring social media could cause Price Modern to lose competitive advantage and market share

leverage. Collective intelligence is the strongest of the collective behavior patterns in the group of companies analyzed, and the results indicate that successful business uses of social media are just emerging. Companies are primarily using blogs, wikis, and social networking—for product delivery, customer service, and brand awareness—to deliver sales and operational effectiveness. Associated patterns for four collective behaviors are shown in Table 10.2. The remaining two behaviors, mass coordination

TABLE 10.4

Strongest Relationship Patterns between a Collective Behavior and the Social Media, Business Use Case, and Business Value

Collective Intelligence	Expertise Location	Interest Cultivation	Relationship Leverage
Associated Collective Behavior			
Expertise location	Collective intelligence	Expertise location	Expertise location
			Interest cultivation
Social Media Most Effectively Used			
Wiki	Social networking	Blog	Blog
Blog			
Business Use Case Most Benefiting			
Product delivery	Customer service	Brand awareness	Brand awareness
	Product delivery		
	Product utilization		
Business Value Most Derived			
Operational effectiveness	Operational effectiveness	Sales effectiveness	Sales effectiveness
	Sales effectiveness		

Source: Bradley, A. (2011). Employing Social Media for Business Impact: Key Collective Behavior Patterns, *Gartner Research*, G00173838, pp. 1–19. With permission.

and emergent structures, are not currently utilized extensively in business environments.

These collective behaviors and their strongest impact patterns can be compared to Price Modern's strategies and strengths to provide a framework for decision making around social media options. All four of the collective behaviors in Table 10.4 closely align with Price Modern's goals. Collective intelligence brings business value to operational effectiveness through improved product delivery and would allow Price Modern to have internal conversations and build upon each employee's experiences. That knowledge could then be collected and disseminated to generate best practice processes. Expertise location can further customer service, a key company value. Interest cultivation strongly corresponds to sales effectiveness and brand awareness, both of which Price Modern seeks in preparing for their launch of a healthcare office furniture initiative. Relationship leverage is also strongly correlated to sales effectiveness and brand awareness which are vital as the Price Modern sales force seeks future business opportunities in a shrinking market.

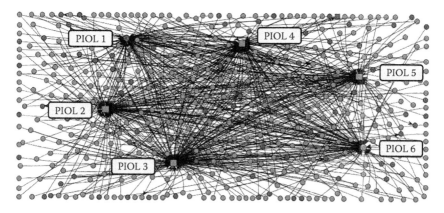

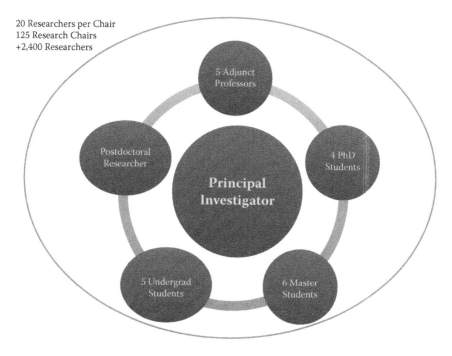

Legend:

- No Speaker Status
- Inactive Speaker
- Active Speaker
- Network Member
- Peer-Identified Opinion Leader (received 2+ nominations)
- Connector (respondent and was nominated)
- Connector + Peer-Identified Opinion Leader

FIGURE 9.5

Example network map: national network map of the top Peer-Identified Opinion Leaders for on-label topics.

20 Researchers per Chair
125 Research Chairs
+2,400 Researchers

5 Adjunct Professors

Postdoctoral Researcher

4 PhD Students

Principal Investigator

5 Undergrad Students

6 Master Students

FIGURE 12.1

The research chair model.

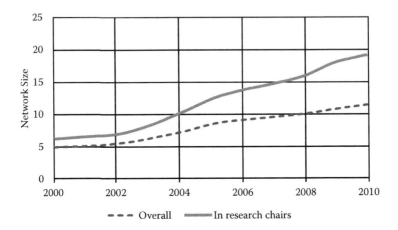

FIGURE 12.2
Trend of collaboration networks size.

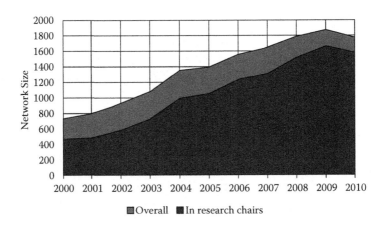

FIGURE 12.3
Trend on scientific production.

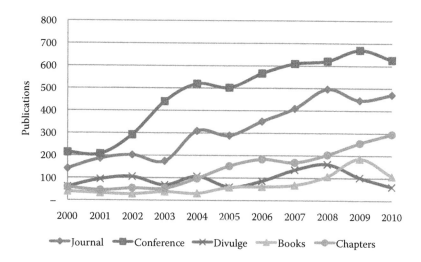

FIGURE 12.4
Research chairs professor publications per type.

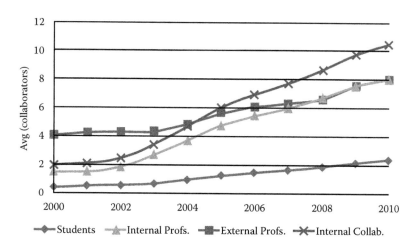

FIGURE 12.5
Type of collaborations in research chairs.

To further define a company's social media strategy, Rozwell (2010) suggests including a definition of the business goals and objectives, the initiatives included in the strategy, and the solutions required to achieve the desired outcomes. In order to assess the potential costs, benefits, and risks associated with the social media strategy, Rozwell (2010) provides three key questions:

1. Which audiences do we want to involve in our social media strategy?
2. What level of engagement do we want with each audience?
3. How much risk can we tolerate in our social media portfolio of initiatives?

The audience for Price Modern includes employees, business partners, customers or prospects, and anyone on the social web (known and unknown). The levels of engagement are defined (Rozwell, 2010) as monitor (organization listens to conversations on the Web), discover (organization analyzes those conversations), share (organization shares its perspective), participate (organization interacts with audience), and cocreate (organization is actively involved with the audience and requests feedback from them). The Price Modern responses to these three questions are

1. Audience: employees, customers, prospects, and influencers
2. Level of engagement:
   a. Employees—create
   b. Customers—discover
   c. Potential customers—discover
   d. Potential healthcare customers—participate
   e. Influencers—discover
3. Risk tolerance: Low

Risk in initiating a social media strategy derives largely from three factors: consistency of the social media strategy with the company's strategic goals, coordination with other business functions, and risk tolerance of the organization. Rozwell (2011) proposes five critical reasons that social media strategies fail:

- Establishing a strategy that is not congruent with enterprise values, strategy, and business goals

- Keeping social media initiatives separated from other traditional business processes
- Embarking on too many uncoordinated efforts
- Using social media to broadcast instead of engage
- Creating social media programs that lack passion and energy and submerge individual creativity

These guidelines formed the design approach to developing a social media strategy for Price Modern.

SOCIAL MEDIA STRATEGIC APPROACH AT PRICE MODERN

Sterne (2010) posits that the "Big Three Goals" of business are increased revenue, lowered costs, and improved customer satisfaction. All of these goals are consistent with Price Modern's business model, and a social media strategy was formulated that addresses some of these goals in the near term and provides a basis for growth in the future within a low risk approach. Hence, Price Modern's social media strategy is a business initiative that strives to

1. Create brand awareness and facilitate communication with the healthcare community related to office furniture in order to become a trusted partner and increase the customer base in that vertical market
2. Increase communication within the Price Modern organization in order to enhance the work environment, share best practices, and increase the ability to better service their customers
3. Enhance the prospecting capabilities of the sales force and streamline their communication

The first objective is related to increased revenue, the second to lowered costs and improved customer satisfaction, and the third to increased revenue and lowered costs. Table 10.5 shows the Price Modern objectives compared to the collective behaviors and most effective social media in Table 10.4.

Price Modern is developing a new Web site for healthcare-related furniture to further the first objective by implementing a customer-centric

TABLE 10.5

Price Modern Objectives and Social Media Implementation (Compare Table 10.3)

	Collective Intelligence	Interest Cultivation	Relationship Leverage	Expertise Location
1. Brand awareness/sales effectiveness		Blog	Blog	
2. Customer service and product delivery/ operational effectiveness	Wiki			Social network
3. Prospecting for clients/ operational effectiveness	Blog			Social network

Source: Bradley, A. (2011), *Employing Social Media for Business Impact: Key Collective Behavior Patterns*, Gartner Research, G00173838, pp. 1–19. With permission.

Web strategy (Landry, 2010). This is a cohesive strategy that involves designing the site in a way that is intuitive to the visitor of that site, putting the customer at the "center of the relationship" consistent with Price Modern's culture. Users of the site were expected to have a variety of backgrounds. Some users were anticipated to be architects and designers who are product savvy and want to be able to quickly locate information, while other users are healthcare workers who may be more familiar with healthcare office terminology and want to have a better understanding of the products that may be available for certain areas of a hospital or laboratory. The addition of a knowledge section where white papers and articles can be posted engages discussions on the latest topics and are value-added features of the Web site for interest cultivation (i.e., Price Modern's desired level of engagement with customers and potential customers—discover). The investment decision in furniture or equipment would be facilitated by relationship leverage through the blog (i.e., level of engagement with healthcare community—participate), and users of the site should be motivated to do business with Price Modern because they have proven that they are knowledgeable and offer value-added services. An analytics tool provides metrics on user patterns such as pages most frequently accessed, topics that inspire the most interest and conversation, and it identifies the audience that visits most often in order to adjust content. Future plans are to add an employee log-in site with additional information that is not public, such as pricing and discounts, so that employees can navigate the site along with their customers, allowing them to provide value as consultants (i.e., level of engagement with employees—create).

The second initiative is to use collective intelligence and expertise location to address Objective 2 with an internal blog and wiki for the best practices program (i.e., level of engagement with employees—create). The blog is called "Price Modern Yammer," and it replaces monthly updates sent by e-mail. Meredith and O'Donnell (2011) propose that there are three classes of social media contributions: contributions of content (primary, secondary, or passive), contribution to the network (such as sending direct messages), and contributions to the platform (such as design or protocols). The Yammer site supports the first two by posting updates and monthly progress reports for comment, allowing employees to voice suggestions and concerns. Approximately 80% of employees joined, and a number of different groups were organized including a "Way to Go" group where people can be publicly thanked or congratulated. There is a best practices group using a wiki to develop and share best practices, and a group to share information about products and issues. The blog and the wiki promoted communication and knowledge sharing. Future plans are to allow people to reach out to help coworkers through some of the struggles that change is likely to bring. There are also plans to use the blog as a way to measure the success or failure of some of the programs and initiatives as well as highlight areas that need further attention.

The third objective was addressed by an initiative to promote prospecting for new clients by utilizing collective intelligence and expertise location. A social network was developed to connect the sales force and management teams (i.e., level of engagement with influencers—discover). To motivate employees to participate, the management team joined as a group, promoted the social network as a virtual place to meet, and encouraged the sales force to participate. The social network is growing in size and impact as salespeople find that it increases their ability to meet others and discover information about projects. Social networking increased lead generation and prospects as people find common interests, and employees are motivated to participate as a way to stay current with the design community and other communities that influence customers and projects. Social networking permits the sales force to be included in various vertical market segments by joining groups associated with those market segments and provides insight into potential customers who might have links to people already known to the company. There is also recruiting potential and the ability to highlight the company to external groups. All three objectives are moving forward in a reasoned, low-risk approach to introducing social media technologies to the organization and its stakeholders.

LESSONS LEARNED

Price Modern's social media strategy was developed from a model that associates collective behaviors with the type of social media, the business use case, and the derived business value (Bradley, 2011). The strategy was refined by considering Price Modern's culture, business goals and objectives, audiences, and levels of engagement and risk (Rozwell, 2010, 2011). The success to date yielded several lessons learned:

- Develop a social media strategy consistent with your company's culture and values.
- Implement the easy wins first to build confidence and prepare the organization for change.
- Determine your company's risk tolerance, and deploy social media strategies that are consistent with that tolerance.
- Develop metrics during the design phase and use them to guide modifications to your social media strategy.
- Understand the threats to success of your social media strategy and develop a mitigation plan.

Price Modern evaluated the primary threat to success of their social media strategy to be lack of use after roll-out, and this assessment influenced the order of implementation of social media. It was decided that acceptance of the internal blog was needed for success of the Best Practices program. Externally, the customer-centered Web site presented the largest potential threat because it required that the company attend to and interact with their users on a regular basis. Recognizing that Price Modern did not have a person dedicated to supporting social media and a customer-centric Web site led to an examination of the resources needed to mitigate risk. Because focus on their customers is central to their corporate strategy, Price Modern will need to ensure that social media is properly supported and of value to their customers.

A key challenge is to convince the sales force that social media will enhance their efforts generating leads, developing and managing relationships, and closing deals. The sales force will need to understand the benefits of sharing information and relationships. Price Modern will need to examine elements such as employee compensation plans to determine if they are rewarding the behavior they desire. Resources are also needed to individually assist employees who are less comfortable with the technology. All of these efforts provide first steps toward a company-wide CRM platform.

FUTURE RESEARCH NEEDS

"CRM on the Cloud" or "social CRM" (Sarner et al., 2010) is the next step in the implementation of a social media strategy for Price Modern. Customer relationships are currently managed individually by each salesperson, some more effectively than others. Providing a means of managing those relationships while allowing sales managers to see progress, marketers to have access to contact information, and a place to house all pertinent information relative to that contact would be a powerful, time-saving tool. It would also give Price Modern a way to gather metrics such as wins and losses, and customer activity over time. Some applications provide even more capability by permitting users to personalize their information and social media to integrate into Price Modern's current portfolio of software.

Price Modern adopted a social media strategy that is consistent with its culture, supported by a change management plan to reduce risk, and oriented to future growth. As technology becomes increasingly social and collaborative, Price Modern is well positioned to benefit from its investment and maintain its market leadership.

ACKNOWLEDGMENTS

The authors would like to thank Price Modern for their encouragement and support of this chapter.

REFERENCES

Austin, T., Drakos, N., Rozwell, C., and Landry, S. (2010). Business Gets Social, Gartner Research, G00207424, pp. 1–7.
Bradley, A. (2011). Employing Social Media for Business Impact: Key Collective Behavior Patterns, Gartner Research, G00173838, pp. 1–19.

Business and Institutional Furniture Manufacturers Association (BIFMA). (2011). The U.S. Office Furniture Market—Historical Growth (Updated March 17, 2011), BIFMA, Grand Rapids, MI. www.bifma.org (accessed March 26, 2011).

Culnan, M.J., McHugh, P.J., and Zubillaga, J.I. (2010). How large U.S. companies can use Twitter and other social media to gain business value, *Management Information Systems Quarterly Executive*, 9(4): 243–259.

Davis, F.D. (1989). Perceived usefulness, perceived ease of use, and user acceptance of information technology, *Management Information Systems Quarterly*, 13(3): 319–340.

FredCavazza.net. (2011). http://gustavofunk.files.wordpress.com/2011/07/socialmedialandscape.jpg (accessed August 12, 2011).

Garrido, L., Cervantes-Pérez, F., González, C., and Mora, M. (2010). Special issue on engineering and management of IDTs for knowledge management systems, *Intelligent Decision Technologies*, 4(1): 1–3.

Kaplan, A.M., and Haenlein, M. (2010). Users of the world, unite! The challenges and opportunities of social media, *Business Horizons*, 53(1): 59–68.

Landry, S. (2010). Hype Cycle for Business Use of Social Technologies, Gartner Research, G00205424, pp. 1–47.

McFarlan, F., Warren, F., McKenney, J., and Pyburn, P. (1983). The information archipelago—Plotting a course, *Harvard Business Review*, 61(1): 145–156.

Melville, N., Kraemer, K., and Gurbaxani, V. (2004). Information technology and organizational performance: An integrative model of IT business value, *Management Information Systems Quarterly*, 28(2): 283–322.

Meredith, R., and O'Donnell, P. (2011). A framework for understanding the role of social media in business intelligence systems, *Journal of Decision Systems*, 20(4): 263–282.

Pearlson, K., and Saunders, C. (2010). *Managing and Using Information Systems*, Hoboken, NJ: Wiley.

Power, D., and Phillips-Wren, G. (2011). Impact of social media and Web 2.0 on decision making, *Journal of Decision Systems*, 20(4): 249–261.

Rozwell, C. (2010). Defining a Social Media Strategy: Identify Audience and Engagement, Gartner Research, G00205700, pp. 1–6.

Rozwell, C. (2011). Avoid Five Critical Failures in Social Media Projects, Gartner Research, G00210398, pp. 1–8.

Rozwell, C., Lapkin, A., and Fletcher, C. (2010). Look Beyond Marketing for Competitive Advantage with Social Media, Gartner Research, G00205916, pp. 1–5.

Sarner, A., Thompson, E., and Maoz, M. (2010). Social CRM Market Definition and Magic Quadrant Criteria, Gartner Research, G00174000, pp. 1–5.

Sterne, J. (2010). *Social Media Metrics*, Hoboken, NJ: Wiley.

Vivacqua, A., James, A., Pino, J., Borges, M., and Shen, W. (2011). Special issue on intelligent collaboration and design, *Expert Systems with Applications*, 38(2): 1077–1078.

11

Visual Knowledge Networks Analytics

Florian Windhager, Michael Smuc, Lukas Zenk,
Paolo Federico, Jürgen Pfeffer, Wolfgang Aigner,
and Silvia Miksch

INTRODUCTION

The potential of social network analysis (SNA) to foster knowledge management initiatives by providing insights into organizational communication and collaboration infrastructures has been the topic of frequent discussion (Swan et al., 1999; Cross et al., 2001, 2003; Marouf, 2007; Cross and Parker, 2004; Marouf and Doreian, 2010). In contrast to the formal tree structures of organizational charts, which are usually created on a management level, the generation of network visualizations is driven primarily from the bottom up by data collected through individual questionnaires or mined from communication databases. The resulting visualizations are able to display vital patterns of informal communication and knowledge sharing, and thus mirror how work actually gets done by teams of actors with different skills and competencies (Krackhardt and Hanson, 1993; Cross et al., 2003). To illustrate some of the insights that SNA methods can offer, Figure 11.1 shows a relational structure (i.e., a node link diagram) that could represent the network of a small organization, obtained by a question like "Who do you ask for advice?"

Networks of this kind can be understood as specific sections through the multilayered fabric of a social unit like a team, group, division, or entire organization. In general, the same social unit exhibits different relational patterns depending on the analytical focus. For instance, a question like "Who asks whom for advice with regard to content X?" will usually show other connections than questions like "Who asks whom for support with

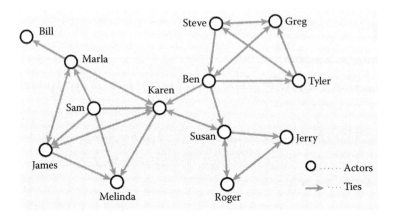

FIGURE 11.1
Example of a possible knowledge communication network for a small organization.

regard to procedure Y?" or "Who is collaborating with whom in current projects?" Given such initial alignments, and accessing the aggregated answers as relational data, SNA methods can now provide a whole spectrum of characteristic measures and insights into the mesh structures of social connectivity (Wasserman and Faust, 1994; Scott and Carrington, 2011).

NETWORK ANALYSIS AND INTERPRETATION

Resulting insights can be classified into two main categories: insights into the shape and structure of macro patterns (like dense clusters of actors and their relative arrangements) and insights into micro network properties like the positions and roles of individual actors in relation to other actors, surrounding clusters, or the whole network.

Visual Network Modeling

Social networks are most commonly modeled as graphs consisting of nodes that represent a set of actors and ties that represent the various relations between these actors. These basic components can carry further information, like individual attributes and labels for nodes, as well as directedness, weight, and further semantic attributes for the ties. For the purpose of representing such graphs visually, ties between nodes are often modeled as elastic springs that pull groups of connected nodes toward each other, while segregating components with less connectivity (di Battista et al., 1999; Krempel, 2005). This layout procedure thus arranges strongly connected actors as (groups of) neighbors, due to a physical force model that tends, technically speaking, to minimize the global stress level of the graph as a whole. Among other effects, such force-directed algorithms locate strongly connected nodes in the center of clusters, whereas sparsely connected nodes are arranged at the margins.

Macro Properties of Networks

As a result of this algorithmic procedure, various properties of a network topology become visible. First glances at network visualizations can provide instant information on whether a social unit is densely connected or not, whether it shows a uniformly distributed shape on the macro level or splits into smaller subgroups, and whether these units are again equally connected within or inhabit dense cores and loosely connected peripheries. The upper left image in Figure 11.2 illustrates such macro-analytical insights by visually highlighting the overall shape and existing subgroups of the example advice network. Driven by the force-directed layout methods outlined above, visual analysis of this data set discloses the existence of three main components separated by structural holes (Burt, 1992) that are only loosely connected by bridging ties between the actors Karen, Ben, and Susan. Revealing macrostructural network properties of this kind fosters a better understanding of the possible barriers facing potential knowledge flows within an organization, as well as insights into the role of single actors as a result of the positions they occupy in relation to each other.

Micro Properties of Networks

Visual analysis on a macro level helps to explore overall communication topologies, and relational micro analysis builds on the awareness that how individuals are situated in a social structure could be of crucial relevance. Networks hence have to be seen as embedding architectures of opportunities and constraints for individual action, and actor-centered measures help to formalize and explore these aspects (Freeman, 1979).

First, crucial information is provided by the specific position of individual nodes within the macro structural topology (i.e., how they are embedded with regard to their distance to central cores, local clusters, or

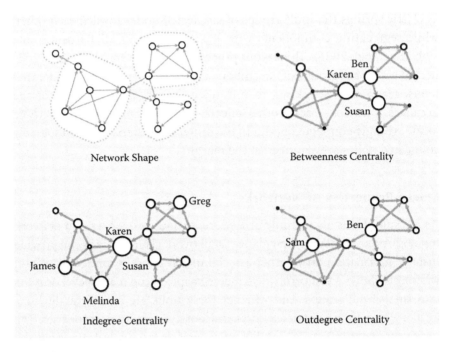

FIGURE 11.2
Visual analysis and various social network analysis (SNA) measures to provide structural insights.

structural holes). The relative position within a topology can be used to derive numerical measures, for instance how central single nodes are with respect to information flows from each node to the others (Koschützki et al., 2005). In the following, we will briefly discuss three different types of centrality measures: betweenness centrality, indegree centrality, and outdegree centrality. Betweenness centrality (Freeman, 1977) calculates the relative importance of actors as transfer stations for communication flows and can be used to identify actors who control communication between various parts of the network. The upper right image in Figure 11.2 visualizes the value of this measure as the relative size of a node. Due to their bridging function, Karen, Ben, and Susan show up with the highest scores of betweenness centrality. For instance, if Karen were taken out of the network, the whole graph would be decomposed into three separated units, whereas the leave of Ben or Susan would cut off their clusters from the overall flow of advice. Another relevant actor measure is the degree (Freeman, 1979) of individual connectivity (i.e., the amount of connections each actor holds). When the directedness of social ties is of relevance (as is the case in advice, power, or distribution networks), degree centrality is differentiated into indegree (all incoming ties) and outdegree (outgoing ties) centrality. The images at the bottom of Figure 11.2 show the values of indegree and outdegree, providing new insights into the role of actors as advice seekers (Sam and Ben with the highest extent) or givers (Karen, Susan, Greg, James, and Melinda with the highest extent).

If degree or betweenness measures of single actors show advanced values, structural key properties are indicated. Key actors with high-degree centrality often serve as hubs or distributors for new threads of communication, whereas nodes with high betweenness can act as gatekeepers for the interaction of adjacent groups or benefit from strategically brokering between different knowledge domains (Burt, 2005).

Knowledge and Change Management Implications

By exploring network properties on the macro or micro level of network configurations, interpretation procedures connect theoretical analysis back to the fields of organizational application and practice. From the knowledge management perspective, insights provided by SNA methods can be of value for various organizational development issues and aims. As Cohen and Prusak (2001) note, knowledge flows along existing pathways in organizations. If we want to understand how to improve the flow of knowledge, we need to understand those pathways. SNA now offers empirical views on the specific architecture of this communication infrastructure. To understand the overall organization or distribution of pathways, all the above-mentioned macro properties and measures are of concern, whereas micro measures help to understand the relational positions, influences, and challenges of single actors. Salient patterns or positions could first be identified, and then further investigated if they meet the expectations in terms of the "should-be states" of an organizational unit. If these analytical results do not meet existing expectations, further investigations could be launched or appropriate knowledge management (KM) or change management (CM) measures initiated in response.

Depending on the particular purpose of an organizational unit, various structures can be expected and deemed appropriate for collectively pursuing intended goals. For instance, a communication structure considered appropriate for tackling iterative production processes might be shown to contain severe barriers for knowledge-intensive research and development projects and vice versa. Depending on the actual situation in different sectors of industry with their different strategy, hierarchy, project, or process configurations—whether actors or units are young and expanding or mature and established—various network structures could do the job of providing appropriate paths for distributing organizational knowledge and expertise, hence facilitating organizational learning cycles.

Given the contextual challenges regarding the question of appropriate network structures, relational analysis alone can rarely offer indisputable interpretations or instant solutions for optimizing network performance. In practice, SNA methods regularly prove to be envisioning procedures that provide a new kind of empirical view on communication and trigger subsequent interpretation processes. These processes could be realized on different levels of organizational participation.

At one end of the spectrum, expert network analysts or consultants from outside the organization collect and analyze data and base their results mainly on their general knowledge of structural patterns and meanings. At the other end of this spectrum, all kinds of participative interpretation can take place, where involved network actors (e.g., employees) combine their rich knowledge about individual and social configurations. They collectively interpret empirical data, generate shared meaning, and create subsequent CM or KM programs. This second setting gives everyone involved, from employees to managers, a temporary role shift, turning them into quasi-organizational analysts to collectively mirror their social behavior and directly deduce interventions.

Knowledge and Change Management Evaluation

The outlined application of network analytical methods to support the alignment and planning of CM and KM programs is already an option for all organizational analysts, because SNA research and software are being distributed and spreading with increasing pace (Huisman and Duijn, 2005; Dandi and Sammarra, 2009). However, these tools rarely include the option to analyze and evaluate initiated interventions and changes with regard to their intended and realized effects.

From a research perspective, this task automatically leads from static to dynamic network analysis, as the differences between (at least) two different chronological instances of the same network have to be identified and compared by organizational analysts. But adding intertime relations to interactor analysis also adds more complexity to the already existing challenges of SNA application in the field. To allow these analyses to also be conducted by nondomain experts, the applied research project ViENA was dedicated to developing methods and software prototypes for visual analytics of dynamic networks.

DYNAMIC NETWORK ANALYSIS

Dynamic network analysis (DNA), which focuses on the evolution, change, and comparison of networks, is a highly active and fast growing field of research (Carley, 2003; Newman et al., 2006), although from a practical perspective, the analytical issues involved are becoming increasingly complex (Bender de Moll & McFarland, 2006; Brandes et al., 2010). The ViENA project (Visual Enterprise Network Analytics) aims to open the use of dynamic network analysis to nondomain experts (e.g., the members of the operations team in an organization). To achieve this, the project addressed several research and development issues.

Initially, DNA procedures had to be selected and bundled into a coherent framework, which would be able to support the real tasks and challenges found in day-to-day business in various organizations. A user and task analysis helped to identify analytical features at the macro and actor levels that support the identification of groups, clusters, and key players. To support the analysis of actual states and the planning and alignment of KM or CM measures, all selected features also had to serve dynamic evaluation purposes, by allowing for before and after comparisons on micro and macro network levels. To lower the thresholds of these complex tasks for nondomain experts, a general strategy of shifting numerical analyses to a visual analytical level was adopted. This would leave computational complexities hidden behind an interactive graphical user interface, thus offering visual overviews, but also showing analytically relevant measures and details on demand. To achieve the former, appropriate visual representation methods for network dynamics had to be found, while the latter required the implementation of additional interaction methods on these representations.

Visual Representations of Network Dynamics

An assessment of the various options available in the developing field of DNA research identified at least four analytical views on network dynamics. These views differ from each other in the way they display different temporal states or panels of the same network. In the simplest case, dynamics (i.e., the display of two network states at time point 1 (t1) and time point 2 (t2), including their structural delta t2 – t1), each view uses different ways of visualizing the dynamics that occur (see Figure 11.3). Layer juxtaposition (Andrews et al., 2009) simply takes networks t1 and t2 and displays them side by side, thus offering analysts a direct comparison. Layer superimposition (Brandes and Corman, 2003) merges the visual structures of t1 and t2, and usually uses colors or additional visual clues to highlight intertime differences (e.g., highlighting emerging or vanishing nodes and ties or tracing the structural shifts of single nodes). A two-point-five-dimensional view (2.5D) (Dwyer and Gallagher, 2004) offers a lateral perspective on a stacked configuration of the network panels t1 and t2, and includes the option of showing intertime traces (trajectories) of nodes, which could carry additional information on network changes. Finally, animation offers moving images that start with state 1 and display a smooth shift of nodes and links toward state 2 (McGrath and Blythe, 2004). Because they each offer different visual advantages and disadvantages (Windhager et al., 2011), a combination of three of these views seemed to offer the best solution for dynamic organizational analyses. This would give the software users the freedom to decide which specific view would best serve their own specific task. In the end, the project team selected the dynamic layer juxtaposition, layer merging, and 2.5D views to offer linked perspectives on changing networks with a user-controlled switching mode based on seamless transitions between them (see Figure 11.3). The fact that the animation method produces artifacts of intrapolation and only allows for the detailed comparison of network states by means of a viewer's memory led to the decision to initially leave it aside.

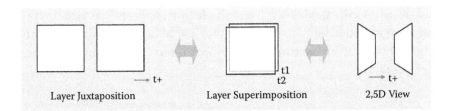

FIGURE 11.3
Three different representations of network dynamics with animated transitions in between to maintain the user's exploration context.

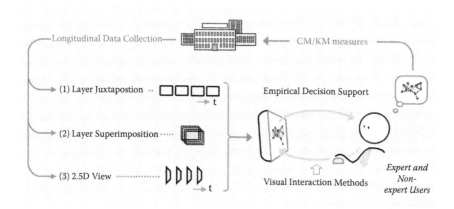

FIGURE 11.4

The functional architecture of ongoing visual knowledge network analysis and organizational self-optimization.

Building on the outlined functions of detailed before-and-after comparison of sequential network states, visual analysis of network dynamics could be extended to ongoing procedures of accompanying organizational knowledge network monitoring. Figure 11.4 illustrates the functional architecture of a corresponding intertwined analysis and change management cycle of visual network analysis, which could be utilized by all interested members of knowledge-intensive organizations to reflect on and optimize their collaborative performance on the basis of empirical network data.

Visual Interaction Methods

Given the three main views on changing network structures (with at least two sequential states in the same network), visual interaction methods had to allow network analysts to extract information that might be of particular interest (e.g., tracking the evolution of node-centered measures over time). To visually navigate and explore a complex network structure, methods like zooming, panning, and rotating help to select appropriate image sections. A linked view architecture assures synchronous visual navigation on parallel network panels (Namata et al., 2007).

Further exploration could now focus on before-and-after comparisons, like the visual investigation of nodes or groups of nodes over time. For instance, to detect whether an actor has gained or lost connections over

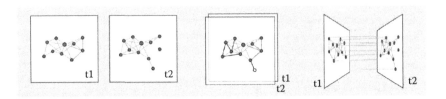

FIGURE 11.5
The display of structural change over time by three different representations.

time, a function that highlights ego-networks helps to analyze change in the layer juxtaposition view. By highlighting the same actor and his or her respective direct contacts on the t1 and t2 panels, organizational analysts can compare whether their cooperative performance increases, decreases, or stabilizes over time (Figure 11.5, left).

Providing organizational analysts with a function that visually tracks the structural position of single nodes within the layer superimposition view allows them to investigate whether a cluster established more or fewer internal connections. Converging tracks of clustered nodes indicate increased cooperation and communication density, whereas diverging tracks signal collaborative decrease, and no visible tracks point toward stability over time (Figure 11.5, center).

The 2.5D view in turn also can display the dynamics of nodes, visualizing them as intertime trajectories between networks t1 and t2, with the angle of these trajectories indicating possible structural shifts. In this view, horizontal alignment indicates stability, whereas inclined trajectories indicate relative moves (Figure 11.5, right). The emergence or disappearance of nodes is made visible by surfacing or vanishing tracks, and when additional node measures (e.g., degree centrality) are mapped onto these trajectories (e.g., by varying track colors), organizational analysts can follow structural and individual changes of performance over time (see Figure 11.8).

To research the opportunities offered by a coherent package of visual analytics techniques for network dynamics with particular regard to its use outside the realm of network experts, the ViENA project bundled all the outlined features, integrated them into a methodical framework, and transferred them into a software prototype (Federico et al., 2011). This implementation paved the way for a conceptual evaluation that took the form of a case study in a knowledge-intensive organization and was designed to show the potential of visual analysis for upcoming network approaches to the dynamic issues of knowledge management, collaborative learning, and organizational development.

CASE STUDY

To ensure our case study had informative potential for the field of applied KM, we chose to investigate a university department (i.e., a particularly knowledge-intensive organization) using a longitudinal network study. Relational data were collected by providing the scientific and administrative staff (33 or 34 employees in each case) with four consecutive online questionnaires over a period of 14 months. Eight different questions addressed the organizational knowledge communication network and included questions on advice relations, intensive project collaboration, need for more communication, and discussion of new ideas to ensure the survey covered the social configurations of innovation. At the start of the case study, the department was structured into three organizational units, with two of these units in the process of merging at the end of the survey. The following example findings were visualized by the ViENA software prototype and optimized for grayscale printing. To preserve the privacy of the organization and its members, the data were made anonymous.

Exemplary Findings

Despite the relatively small size of the target organization, the dynamic data set with its eight different layers of knowledge communication at four chronological panels opens up a wide variety of visual analysis options, especially the use of different views and various interaction methods. The selected visualizations have thus only been chosen to provide an impression of the actual possibilities that are available.

Starting with the layer juxtaposition view, the evolution of a selected network becomes visible sequentially from left to right. Figure 11.6 shows a close-up view of a network that resulted from the question "With whom do you think you should communicate more often to achieve your intended working objectives?" The resulting structure shows relations that could be classified as an organizational development plan for communication, put together from the bottom up by all employees from their practical daily work perspective. The left side of the image shows the first of four network states, with the same network 3 months later shown on the

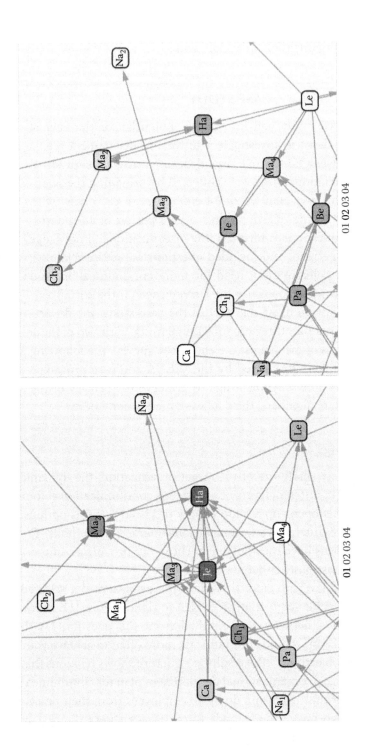

FIGURE 11.6

Layer juxtaposition view on a network representing relations where more communication or collaboration was required to optimize achievement of individual working objectives.

right. The nodes are colored by their in-degree centrality (computing how often an actor was chosen by others), showing those with a high value in dark gray (i.e., classified as in high demand for more communication), and those considered adequately reachable by all other actors in light gray.

If we focus on the configuration of ties, and take the node colors into account, we can see that two actors in particular—Jeff (Je) and Hans (Ha)—show a high in-degree at network t1 (left) (i.e., are in high demand for more communication from various sides). A subsequent organizational restructuring process that defined new formats of frequent exchange and clarified organizational responsibilities sought to react to these needs. Three months later, the communication demand network t2 shows a notable improvement, revealing lesser ties indicating demand for Jeff and Hans, as well as a reduced overall density of the visualized need for more communication network.

If we switch from the outlined juxtaposition perspective to the layer superimposition view, a new kind of structural change becomes visible. Figure 11.7 shows the organizational innovation network that resulted from the responses to the question "With whom do you discuss new ideas?" All four sequential states of this network are superimposed, and only the relative shifts of individual node positions (instead of the innovation relations) are shown by arrows from t1 to t4. Nodes are colored according to organizational unit. The overview on the left is complemented by a close-up view on the members of the white unit on the right.

Depending on the evolution of their relational embeddedness, the visualized tracks of single nodes show their relative paths toward the center or the periphery of a given cluster. Taken together, all tracks show the global trends in the evolution of the network. Converging tracks indicate increasing exchange of new ideas, and diverging tracks show decreasing communication regarding new ideas. Following this global perspective, the overview on the left panel in Figure 11.7 shows a relative approximation of the dark gray cluster at the top, the twofold cluster at the bottom, including the light gray, as well as a large part of the white unit. In turn, the close-up on the right reveals a significant visual divergence in the innovation sharing network. As a result of an intended merger of the white and light gray units and associated progressive reorientation process of the actors Nadja (Na), Judith (Ju), Jack (Ja), Ines (In), and Jeff (Je), the superimposition view shows the final separation of these nodes from the white unit's innovation network, which formally merged with the light gray unit between t3 and t4.

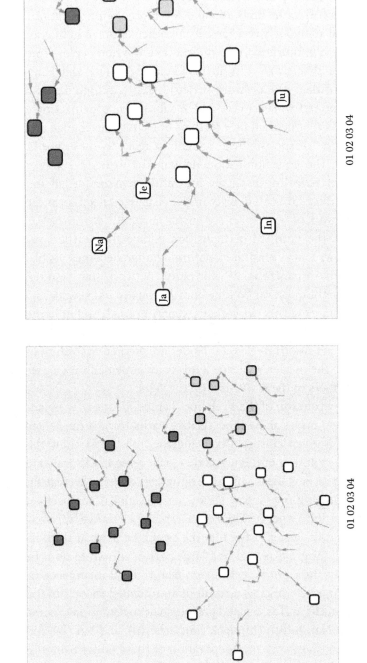

FIGURE 11.7

Layer superimposition view (overview and detail) on the organizational innovation network with arrows showing the shift of individual node position from t1 to t4.

While the layer superimposition view is already able to visualize important changes like the hiring and departure of nodes using stepwise reduced chains of arrows, the 2.5D view on network change allows for an explicit perspective on such transitions and the fine-grained evolution of selected node values. Figure 11.8 shows the network of those actors who provided names in response to the question: "Who covers your areas of expertise when you are not there?"

When all nodes and their trajectory bundles are shown in a 2.5D perspective, similar analytical functions to the superimposition view (e.g., detecting trends in individual or global structural shift) can be exploited. The clearly visible trajectories in 2.5D add the option to precisely detect the beginning and end, as well as the evolution of various actor values, which could be mapped on to these traces using additional colors. In Figure 11.8, only trajectories of those nodes that leave or join the network between t1 (left) and t4 (right) are shown. Mapped onto these visual traces are their respective in-degree values (dark color indicates high in-degree) (i.e., how often they were named as substitute for a certain area of expertise).

From a knowledge management perspective, this view uncovers important knowledge carriers and helps to identify key actors who stabilize the knowledge network, and thus should not be lost without transferring as much of their expertise as possible to their neighboring or succeeding actors. Against this background, Figure 11.8 shows that the intended replacement of Stan (St) and Nadine (Na) who are leaving at t3 could not

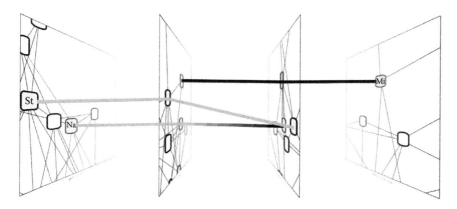

FIGURE 11.8

Two-point-five-dimensional (2.5D) view of a close-up of the knowledge substitution network (time line from left to right) with activated trajectories showing actors who had left (St and Na) or joined the organization (Mi).

be fully covered by the new hire Michael (Mi). The two actors who are leaving were frequently named as knowledge substitutes (high in-degree, visualized as light gray trajectories), while the new actor is not nominated to a comparable extent. As a reaction to this evaluation, specific training was initiated for the people concerned and a second vacancy advertised.

To summarize, when processed by a new software prototype for dynamic network visualization, the case study data allowed the organization to explore its knowledge sharing and communication structures in a new way. The planning and evaluation of various KM and CM measures were supported with empirical findings and will now be extended to future self-evaluation and visual knowledge network monitoring cycles.

SUGGESTIONS FOR FURTHER RESEARCH

Several relevant research and implementation issues emerged with regard to extended application in the field of knowledge management. As far as data collection is concerned, knowledge network analysis faces major challenges from the privacy and data protection points of view. Even when the highest standards are extended and applied to all participants in an organizational network survey, substantial privacy concerns have to be expected from various sides. The only way to cope with these challenges would seem to be to employ a fully transparent research procedure that takes account of concerns on all organizational levels and thus already assures a high level of commitment in the run-up to a network study. Complementing efforts toward participative interpretation cycles could help to clarify the benefits of gaining insights into collaborative structures and new ways of working on their progressive enhancement.

Developments in these aspects could also lay the foundations for complementary developments on the technical sides of data collection and visualization, which tend to come up against various barriers when facing dynamic large-scale networks and organizations. To cope with vast amounts of nodes, the development of dynamic hierarchical clustering methods has to gain increased attention (Parthasarathy et al., 2011). To bring individual knowledge back into network analysis, methods of integrated attribute and structure visualization have to be enhanced and matured (Ibarra et al., 2005; Krempel, 2005). Only when data on the existing knowledge of individual actors can be joined with knowledge of collective knowledge sharing infrastructures can valuable analyses of distribution specifics of particular knowledge domains be mapped and evaluated.

Another new frontier in organizational analysis is the integration of organizational process data into enriched analytical computation and visualization environments (Blickle and Heß, 2006; Windhager et al., 2011). Merging these two paradigms would allow insights into the systematic interplay of sequential performance (in terms of chronological chains of individual learning and action) and their structural interweaving via communication and collaboration networks. To achieve this, the visualization-oriented network and process modeling research communities should consider starting to connect their own complementary knowledge domains to develop new ways of providing insights for embedded actors in constantly evolving organizational knowledge networks.

CONCLUSIONS

As discussed, social network analysis can be of high relevance for supporting knowledge management initiatives at different stages of their implementation. First, the empirical analysis of existing pathways and barriers to knowledge communication on a macro level can help to generate a deeper understanding of a target network configuration. Second, the investigation of micro network properties helps to understand the role of single (key) actors within the overall structure of an organization. Investigations and insights on both levels will enable people in organizations to plan and align knowledge and change management measures based directly on empirical data. For instance, while macro level analysis can help to develop communication strategies between different organizational units, micro level analysis can help to identify the key actors who need to be addressed to effectively implement these strategies through KM or CM initiatives. This one-time network analysis option already provides strong planning and decision support, but we contend that dynamic network analysis offers major value added through the additional opportunity it offers to evaluate the results of KM and CM initiatives on an empirical level. To also allow people with no specific SNA expertise to perform these analyses, we presented a framework that shifts analytical procedures from a computational to a visual level, adding further visual interaction methods for detailed analysis and ongoing network monitoring. As shown by our longitudinal case study of the knowledge communication network of a university department, this approach enables organizations to reflect, support, and evaluate their day-to-day knowledge communication on various levels of granularity and with regard to the specific challenges and questions that guide their collective work. Consequently, we see the main contribution of SNA to knowledge management to lie in its ability to replace educated guesses on the intended effects of various interventions with evidence obtained from empirical insights into a given structure and its actual evolution. Analogous to the benefits of statistical diagrams, we consider dynamic network visualizations to deliver valuable overviews of highly complex temporal organizational data. By enriching these overviews with visual interaction methods, analysts, managers, and employees can also be given access to detailed exploration options. Further research therefore has to address the question of how they could be provided with tools that are comparable to and similar to use as other information technologies already in use in the modern organizational context.

ACKNOWLEDGMENTS

This research was supported by the Austrian FFG research program FIT-IT Visual Computing (Research project ViENA, Visual Enterprise Network Analytics, http://fitit-viena.org) as well as by the Centre for Visual Analytics Science and Technology (CVAST), funded by the Austrian Federal Ministry of Economy, Family, and Youth under the exceptional Laura Bassi Centres of Excellence initiative.

REFERENCES

Andrews, K., Wohlfahrt, M., and Wurzinger, G. (2009). Visual graph comparison. *Proceedings of the 13th International Conference on Information Visualization* (IV09) (pp. 62–67). Los Alamitos, CA: IEEE Computer Society.

Bender-deMoll, S., and McFarland, D.A. (2006). The art and science of dynamic network visualization. *Journal of Social Structure*, 7(2). er 2.

Blickle, T., and Heß, H. (2006). From process efficiency to organizational performance. In A.W. Scheer, H. Kruppke, W. Jost, and H. Kindermann (Eds.), *Yearbook Business Process Excellence 2006/2007* (pp. 269–281). Berlin: Springer.

Brandes, U., and Corman, S.R. (2003). Visual unrolling of the network evolution and the analysis of dynamic discourse, *Information Visualization*, 2(1): 40–50.

Brandes, U., Indlekofer, N., and Mader, M. (2011). Visualization methods for longitudinal social networks and actor-based modeling, *Social Networks* (preprint submitted June).

Burt, R.S. (1992). *Structural Holes: The Social Structure of Competition*. Cambridge, MA: Harvard University Press.

Burt, R.S. (2005). *Brokerage and Closure: An Introduction to Social Capital*. New York: Oxford University Press.

Carley, K. (2003). Dynamic network analysis. In R. Breiger, K. Carley, and P. Pattison (Eds.), *Dynamic Social Network Modeling and Analysis: Workshop Summary and Papers* (pp. 133–145). Washington, DC: National Academies Press.

Cohen, D., and Prusak, L. (2001). *In Good Company: How Social Capital Makes Organizations Work*, Cambridge, MA: Harvard Business School Press.

Cross, R., Parker, A., Prusak, L., and Borgatti, S.P. (2001). Knowing what we know: Supporting knowledge-intensive work. *Organizational Dynamics*, 30: 100–120.

Cross, R., Parker, A., and Sasson, L. (2003). Introduction. In R. Cross, A. Parker, and L. Sasson (Eds.), *Networks in the Knowledge Economy*. New York: Oxford University Press.

Cross, R., and Parker, A. (2004). *The Hidden Power of Social Networks: Understanding How Work Really Gets Done in Organizations*. Boston, MA: Harvard Business School Press.

Dandi, R., and Sammarra, A. (2009). Social network analysis: A new perspective for the post-fordist organization. *Proceedings from ASNA '09: Sixth Conference on Applications of Social Network Analysis*. Zürich, Switzerland.

di Battista, G., Eades, P., Tamassia, R., and Tollis, I.G. (1999). *Graph Drawing: Algorithms for the Visualization of Graphs*. Englewood Cliffs, NJ: Prentice Hall.

Dwyer, T., and Gallagher, D.R. (2004). Visualising changes in fund manager holdings in two and a half-dimensions. *Information Visualization*, 3(4): 227–244.

Federico, P., Aigner, W., Miksch, S., Windhager, F., and Zenk, L. (2011). A visual analytics approach to dynamic social networks. In S. Lindstaedt and M. Granitzer (Eds.), *Proceedings of the 11th International Conference on Knowledge Management and Knowledge Technologies (i-KNOW '11)* New York: ACM Press.

Freeman, L.C. (1977). A set of measures of centrality based on betweenness, *Sociometry*, 40: 35–41.

Freeman, L.C. (1979). Centrality in social networks: Conceptual clarification. *Social Networks*, 1(3): 215–239.

Huisman, M., and Van Duijn, M.A.J. (2005). Software for social network analysis. In P.J. Carrington, J. Scott, and S. Wasserman (Eds.), *Models and Methods in Social Network Analysis* (pp. 270–316). New York: Cambridge University Press.

Ibarra, H., Kilduff, M., and Tsai, W. (2005). Zooming in and out: Connecting individuals and collectivities at the frontiers of organizational network research. *Organization Science*, 16(4): 359–371.

Koschützki, D., Lehmann, K.A., Peeters, L., Richter, S., Tenfelde-Podehl, D., and Zlotowski, O. (2005). Centrality indices. In Brandes, U., and Erlebach, T. (Eds.), *Network Analysis: Methodological Foundations* (pp. 16–61), LNCS 3418, New York: Springer.

Krackhardt, D., and Hanson, J.R. (1993). Informal networks: The company behind the chart. *Harvard Business Review*, 71(4): 104–111.

Krempel, L. (2005). *Visualisierung komplexer Strukturen: Grundlagen der Darstellung mehrdimensionaler Netzwerke*. Frankfurt am Main: Campus.

Marouf, L. (2007). Social networks and knowledge sharing in organizations: A case study. *Journal of Knowledge Management*, 11(6): 110–125.

Marouf, L. and Doreian, P. (2010) Understanding information and knowledge flows as network processes in an oil company. *Journal of Information & Knowledge Management,* 09: 105–118.

McGrath, C.J., and Blythe, J. (2004). Do you see what I want you to see? The effects of motion and spatial layout on viewers' perceptions of graph structure. *Journal of Social Structure,* 5(2).

Namata, G.M., Staats, B., Getoor, L., and Shneiderman, B. (2007). A dual-view approach to interactive network visualization. In *Proceedings of the Sixteenth ACM Conference on Information and Knowledge Management* (pp. 9393–9942). New York: ACM Press.

Newman, M.E.J., Barabási, A.L., and Watts, D.J. (2006). *The Structure and Dynamics of Networks.* Princeton, NJ: Princeton University Press.

Parthasarathy, S., Ruan, Y., and Satuluri, V. (2011). Community discovery in social networks: Applications, methods, and emerging trends. In C. Aggarwal (Ed.), *Social Network Data Analytics.* New York: Springer.

Scott, J.P., and Carrington, P. (Eds.). (2011). *The SAGE Handbook of Social Network Analysis.* Thousand Oaks, CA: Sage.

Swan, J., Newell, S., Scarbrough, H., and Hislop, D. (1999). Knowledge management and innovation: Networks and networking. *Journal of Knowledge Management,* 3(4): 262–275.

Wasserman, S., and Faust, K. (1994). *Social Network Analysis: Methods and Applications.* Cambridge, MA: Cambridge University Press.

Windhager, F., Zenk, L., and Federico, P. (2011, in press). Visual enterprise network analytics—Visualizing organizational change. *Procedia—Social and Behavioral Sciences,* Elsevier, 58–67.

12

A Framework for Fostering Multidisciplinary Research Collaboration and Scientific Networking within University Environs

Francisco J. Cantú and Héctor G. Ceballos

INTRODUCTION

Socializing and collaboration are in the very nature of human beings. Aristotle, in the fourth century B.C. defines humans as a kind of "social animal" identifying socializing as a core attribute of the entity humans, posing this attribute at the same level as reasoning the *sine qua non* feature of humankind. Since early ages, humans organized and gathered together to cooperate in order to survive and assure the continuation of the species, and we observe this behavior in the various eras of history as well as in modern times. With the advent of experimental science, the industrial revolution, and the revival of Kantian philosophy and positivism in the XIX century, we witness the emergence of social sciences and in particular, sociology, led by scholars such as Émile Durkheim, Ferdinand Tönnies, and Georg Simmel, who studied social phenomena from a philosophical and scientific standpoint introducing concepts and background theory, and contributing in this way to the establishment of social science as a discipline by the end of the nineteenth century and laying down the concepts for what is now known as social networks. In the first half of the twentieth century different approaches to social networks theory and practice were developed in the United States and the United Kingdom. With the advent of computers in the 1960s and 1970s and as a result of the work of a growing number of social science scholars, new methods and analytical tools for social networks appeared in various universities including Harvard, California, Chicago, and others. In parallel, advances in computer science and electronics led to the establishment of information and communication technologies in the last few decades, which have become enablers for new means of communication and collaboration among persons and groups in contemporary society.

Research during the 1970s in the Defense Advanced Research Projects Agency (DARPA) project laid down the foundations for the establishment of the Internet during the 1980s and means of communication such as electronic mail and interactive chatting sessions among remote individuals. The growth of public electronic sites in the 1990s was supported by tools such as *Mosaic* and others, which became predecessors of the *World Wide Web*. Navigators such as *Netscape*™ and *Explorer*™ also played an important role in facilitating virtual traveling through the Web of sites around the world. The need for searching mechanisms that would assist users in finding useful information among the myriads of data stored in millions of Web sites around the world soon became evident, and the solutions appeared without delay with the invention of search engines such as *Google*™, *Yahoo*™, and others with built-in intelligence implemented in sophisticated computer algorithms. Finally, to satisfy the inherent need of humans to communicate and socialize, theories and methods for social networking were conceived to play such a function, leading to the inventions of tools such as *Facebook*™ and *Twitter*™, among others (Sudeshna, 2010).

Thus, the world is communicated in various ways, and people are using such means to share values, faith, beliefs, and hopes, in organizing themselves for achieving aspirations based on values of freedom, justice, and fraternity, such as recent events in North Africa and the Middle East have shown. Nations collaborating for exchange and improvement in economic development is another social networking with initiatives such as the one from the Organization for Economic Cooperation and Development (OECD) and other initiatives (The Royal Society, 2011). What happens at national or regional levels may also be mirrored at institutional levels when organizational culture, traditions, and practices may not adapt as quickly as needed to respond appropriately to new challenges in business, public management, and education in a changing and communicated world.

We are well communicated and networked with external parties, but the same is not necessarily true when we look inside our organization and realize that we do not know what my officemate, factory partner, or academic department colleague does. I may know well what a colleague in my same discipline who lives on another continent is working on, but ignore what projects my university colleagues from other departments are doing. Concepts such as the intranet and extranets were developed to notice and become conscious of the need for internal communication, collaboration, and socialization. In this chapter, we address the importance of inner collaboration and social networking within an institution to better achieve their goals and objectives and present a model and experience in fostering inner collaboration and networking among research groups from multiple disciplines in a university environment (Byrne, 2010).

BACKGROUND

Social networks theory, practice, and tools developed particularly in the last few decades, including what is known as *social network analysis* (SNA). SNA is a formal approach for the study of social networks using theory, methods, and tools from *Graph Theory* for the mathematical analysis of social processes. It uses graph theory, pattern recognition, and data mining methods to discover and measure properties and patterns on graphs and networks. A network is stated as a collection of entities and their interactions in which entities are called *nodes* and interactions are called *arcs* or *edges*. Nodes represent individuals, classrooms, workplaces, families, countries, and other kinds of entities. The arcs or edges represent interactions or relationship between nodes of the network. Properties of networks include concepts such as *centrality* (how well connected is a node with respect to other nodes), *betweenness* (how well a node connects sets of nodes), *closeness* (how distant a node is with respect to other nodes), and some others (Wasserman and Faust, 1994).

SNA software tools assist in doing quantitative or qualitative analysis of social networks by finding properties of a network which are shown either as tables of numerical attributes or by displaying visual representations. Among popular SNA software tools are *C-IKNOW*, a Web-based software tool for numerical and visualization analysis; *Commetrix*, a framework for dynamic network analysis and visualization that is applied to study coauthorship, e-mail, and newsgroups; UCI-Netdraw; and *CFinder*, for finding and visualizing overlapping dense communities in social networks using the clique percolation method (Freeman, 2006).

SNA was applied to study patterns of collaboration in scientific fields including physics and other disciplines (Newman, 2001). Newman analyzes the structure of scientific collaboration networks in terms of coauthorship and shows that any randomly chosen pairs of scientists are typically separated by only a short path of intermediate colleagues. Newman et al. (2002) demonstrate the presence of clusters in networks and the patterns of collaboration between the scientific fields (Newman et al., 2002; Newman, 2004). David Liben-Nowell and Jon Kleinberg propose the *link-prediction problem* to infer new interactions among members of a social network which are likely to occur in the near future based on measures of proximity of nodes and present experiments on large publications networks to predict future coauthorship (Liben-Nowell and Kleinberg, 2007).

In the following section, we present a model and case study in research collaboration and multidisciplinary scientific networking in a university in which we apply some of the principles of SNA and scientific collaboration following Newman´s concepts and methods for establishing patterns of behavior in networks of scientific research and collaboration.

A CASE STUDY IN RESEARCH COLLABORATION AND SCIENTIFIC NETWORKING

We now present a case study in research collaboration and multidisciplinary scientific networking conducted at Tecnológico de Monterrey, a comprehensive teaching and research university located in Monterrey, Mexico. The academic staff is composed of around 800 research professors and around 1,600 research students at the doctoral and master level from disciplines in engineering, information technologies, social sciences, arts and humanities, natural sciences, and health sciences. This conglomerate of 2,400 researchers are organized in groups called *research chairs* with a principal investigator as the group leader as well as various adjunct professors and graduate students for an average of 20 researchers per chair. Thus, a research chair is a kind of collaborative scientific social network integrated by multidisciplinary researchers. For instance, the research chair in medical engineering congregates researchers from mechanical engineering, electrical engineering, and medicine. A research chair in border studies gathers researchers from economics, demography, and international relations. Similarly, a research chair in student learning

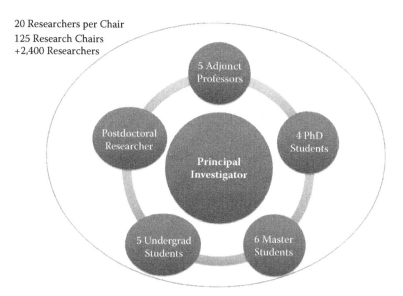

20 Researchers per Chair
125 Research Chairs
+2,400 Researchers

FIGURE 12.1 (See color insert.)
The research chair model.

integrates researchers from education, information technologies, and psychology. Today, there are 125 research chairs in the various disciplines that followed the model depicted in Figure 12.1 since 2003 when they started (Cantú et al., 2009).

A NETWORKING AND COLLABORATION DATABASE

Scientific publications from research chair members are registered in a database administered by the university. An information system called *SIIP* (*Sistema de Información para la Investigación y el Posgrado* [*Information System for Research and Graduate Programs*]) was implemented in 2004 to register their scientific publications including journal articles, conference papers, patents, books and book chapters, theses, and technical reports. It also stores information about research professors and graduate students, research chairs, research centers, and graduate programs in a relational database. When registered, publications are associated explicitly to a research chair by their respective authors. A publication can be associated to several research chairs as long as at least one of its authors belongs to that chair. Sharing credit in publication stimulated collaboration among groups and attained acquaintance with the work of groups in all the disciplines. SIIP is a type of multiagent system with learning capabilities and data mining facilities as described in Cantú and Ceballos (2010). Studies to analyze social behavior for networking and collaboration over publications of Tecnológico de Monterrey researchers were conducted by Valerio and collaborators (Treviño et al., 2007).

Research professors' academic profile can be obtained by any internal user from the SIIP database in the form of a short curriculum vitae or as a full one with the main scientific publications organized by journal, conference, patents, books, theses, technical reports, and other products. Professor expertise is given in terms of keywords and industrial sectors, which narrows the search of specialists in certain areas. Professors also register scientific stays on other universities, indicating the visited department and a contact on the other university. This information is used for developing a catalog of universities with which there exists collaboration agreements. This catalog is used for quantifying the level of exchange and collaboration with these universities. SIIP classifies the list of participants of research chairs in three categories: internal researchers, students, and foreign researchers. The profile of internal researchers is used for characterizing the specialization of the research group. In the case of students and foreign researchers, their profile is more limited. For students, we import information like their background, the program in which they are enrolled, thesis advisor, and dissertation topic. For foreign researchers, the university and department they belong to are registered in the system. This information along with publications registered in the SIIP allow for characterizing the research group and its specialization area which must be aligned with one of the strategic areas of the Tecnológico de Monterrey. Coauthorship in publications allows for measuring the level of participation of students and foreign researchers in the group. In the opposite direction, we also measure the relationship between graduate programs and research chairs based on the number of master and PhD theses aligned with some line of the research chair. This alignment is made explicit when the thesis is registered.

A GUIDE FOR STUDENTS

Prospective and enrolled students may consult the profile of professors attached to a graduate program in order to select a thesis advisor or reviewer. Professors' profiles include scientific production, thesis advised, patents, expertise (given in keywords and industrial sectors), research projects, and research chairs. This information is used for identifying other professors with similar interests or developments. Related professors are ranked based primarily on their keywords, journals in which they publish, and collaboration on research chairs and projects. Students additionally must join a research chair. In order to select one aligned with their interests they can browse the catalog of research chairs by area, specialty, participants, projects, publications, and so forth. Students can address chairs or participants directly through the system. Students can also identify universities with whom Tec de Monterrey has collaboration in order to plan a research stay or choosing an external advisor. This collaboration is given by coauthorship on publications, research stays, and research projects. These elements are used for calculating an index of collaboration that in turn is used for ranking universities in certain areas.

COLLABORATION NETWORKS

In order to measure the impact of research chairs on the collaboration of our professors, we used the concept of *collaboration network* proposed by Newman (2004). This refers to the list of distinct coauthors of publications in the last 5 years for a given professor. We observed a constant increment on both scientific production and collaboration among researchers since 2003 when the research chairs program started. We observed that the average collaboration network size passed from 5 in the year 2000 to 11.5 in the year 2010, where the collaboration network size is the average number of coauthors for a given publication. It is worth noticing that in the case of professors participating in research chairs, the increment on the network size passed from 6.3 to 19.1 in the same period. That is to say, the research chair position creates collaboration by its very nature. This pattern of behavior is illustrated in Figure 12.2.

Similarly, the number of journal and conference researcher publications passed from 729 in the year 2000 to 1,779 in the year 2010. This increment is mostly attributed to professors participating in research chairs, as can be seen in Figure 12.3.

Figure 12.4 shows the variations in different types of publications considered in this study: articles in journals, articles presented at conferences,

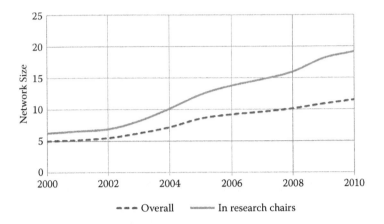

FIGURE 12.2 (See color insert.)
Trend of collaboration networks size.

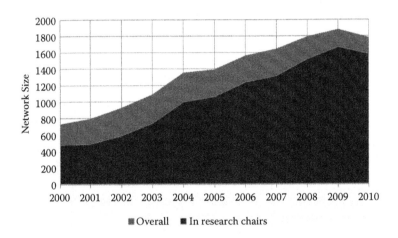

FIGURE 12.3 (See color insert.)
Trend on scientific production.

articles in divulge journals, books, and book chapters. As can be seen, the number of books and articles in divulge journals fluctuates on the same range, meanwhile the number of articles in refereed journals, articles presented at conferences, and book chapters show a relatively constant increment.

On the other hand, we distinguish three types of collaborations in publications: with students, with professors of our university, and with professors at other universities. The first two represent internal collaboration, whereas the last represents external collaboration. Figure 12.5 shows the

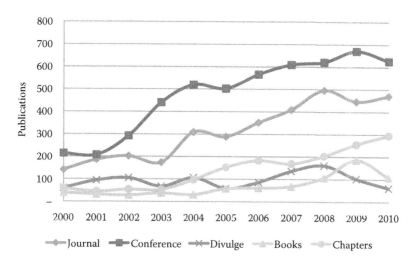

FIGURE 12.4 (See color insert.)
Research chairs professor publications per type.

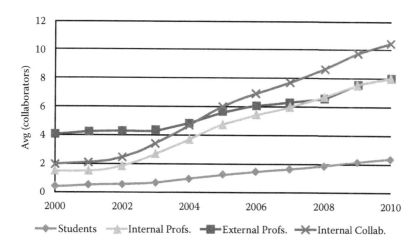

FIGURE 12.5 (See color insert.)
Type of collaborations in research chairs.

number of collaborations with students, colleagues of the same university, and colleagues of other universities on average for professors participating in research chairs, as well as the total internal collaboration (the sum of the first two). As can be seen in Figure 12.5, during earlier years external collaboration prevailed over internal collaboration. Nevertheless, in 2003 internal collaboration increased drastically, passing from 3.5 to 10.5 in 2010, even surpassing external collaboration. Growth of internal collaboration has the same pattern for both students and university colleagues. Finally we can observe that for 2010 the collaboration with colleagues of the same university and from other universities is practically the same (McDonald, 2003).

MULTIDISCIPLINARY RESEARCH NETWORKS

By design, a research chair gathers researchers from various disciplines. If collaboration among researchers really happens, this must be reflected in the number of coauthored publications within a research chair as well as among research chairs. We analyzed both parameters and found that many of the chairs have at least one publication with authors from at least two disciplines and that a good number of the publications have one or more authors from at least two research chairs.

CONCLUSIONS AND LESSONS LEARNED

We presented a model for multidisciplinary collaborative research and scientific networking within a university based on the concept of a research chair and presented evidence that this model is fostering collaboration in the form of coauthorship between the various types of researchers within a chair and between disciplines. The support of an intelligent multiagent system called SIIP proved useful in providing a mechanism to record scientific publications and in attaining acquaintance on the work of other researchers within the university. Thus SIIP stimulated coauthorship among professors and graduate students of various disciplines. SIIP has other uses that were not mentioned here such as the distribution of performance indicators, the automatic distribution of alerts and notifications to authors, the quality of publications according to citations and impact factor, and the periodic evaluation of research.

FUTURE WORK

Further studies along the ideas outlined by Newman, Liben-Nowell, Kleinberg, and others will be conducted in the near future on the SIIP database to learn more about patterns of publications, correlations of various types, inner and external connectivity, and distance on network paths of authorship among researchers and disciplines (Chincilla-Rodriguez et al., 2008).

REFERENCES

Byrne, Tony. (2010). How to use internal collaboration and social networking technology. *Sales and Marketing Newsletter*, INC.Com. p. 209.

Cantú, F., Bustani, A., Molina, A., and Moreira, H. (2009). A knowledge-based development model: The research chairs strategy. *Journal of Knowledge Management*, 13(1): 154–170.

Cantú, F., and Ceballos, H. (2010). A multiagent knowledge and information network approach for managing research assets. *Expert Systems with Applications; An International Journal*, 37: 5272–5284.

Chinchilla-Rodríguez, Z., de Moya-Anegón, F., Vargas-Quesada, B., Corera-Álvarez, E., and Hassan-Montero, Y. (2008). Inter-institutional scientific collaboration: An approach from social network analysis. *Proceedings of the Prime-Latin America Conference*, G. Dutrenit, Ed. Mexico City: UAM-UNAM. pp. 1–24.

Freeman, Linton. (2006). *The Development of Social Network Analysis*. Vancouver: Empirical Press.

Liben-Nowell, David, and Kleinberg, Jon. (2007). The link-prediction problem for social networks. *Journal of the American Society for Information Science and Technology*, 58(7): 1019–1031.

McDonald, David W. (2003). Recommending collaboration with social networks: A comparative evaluation. *Proceedings of the ACM Conference on Human Factors in Computing Systems*. G. Cockton and P. Korhonen, Eds., Fort Lauderdale, FL.

Newman, M.E.J. (2001). Scientific collaboration networks: I. Network construction and fundamental results. *Journal of Physical Review E, American Physical Society*, 64(1): 016131.

Newman, M.E.J., Watts, D.J., and Strogatz, S.H. (2002). Random graph models of social networks. *Proceedings of the National Academy of Sciences of the USA*, 99(1): 2566–2572.

Newman, M.E.J. (2004). Coauthorship networks and patterns of scientific collaboration. *Proceedings of the National Academy of Sciences of USA*, 101: 5200–5205.

Sudeshna, D., Rogan, M., Kawadler, H., Brin, S., Corlosquet, S., and Clark, T. (2010). PD Online: A case study in scientific collaboration on the Web. *Proceedings of the FWCS*, E. Prudhommeaux, Ed. Raleigh, NC, 1–7.

The Royal Society. (2011). *Knowledge, Networks, and Nations: Global Scientific Collaboration in the 21st Century*, London.

Treviño, Ana, Valerio, Gabriel, and Ramírez, Pablo. (2007). Social knowledge networks at ITESM, *Proceedings of the Fifth International Conference on Formal Concept Analysis*. S. Kuznetsou and S. Schmidt. Clemont-Ferrand, Eds. France: Springer. pp. 25–32.

Wasserman, S., and Faust, K. (1994). *Social Networks Analysis: Methods and Applications*. Cambridge: Cambridge University Press.

13

Knowledge Management and Collaboration: Big Budget Results in a Low Budget World

Andrew Campbell and Melvin Brown II

It is a new version of an old story. Companies like Microsoft developed and deployed powerful collaboration platforms that are massively under-utilized because they are viewed and implemented (i.e., deployed) as traditional software. Moreover, several of these platforms are on their third or fourth iteration and are robust by any measure. Some, like Microsoft® SharePoint®, were purchased and incorporated into organizational software portfolios at a level approaching ubiquity.

The second phase of this shift has not really happened yet. With a few exceptions, these platforms (Gartner's category is Social Software in the Workplace—Magic Quadrant for Social Software in the Workplace, 25 August 2011, Drakos, Mann, Rozwell, and Sarner, Research Note G00214779) were purchased and installed by information technology (IT) shops in organizations and have fallen far short of their potential. Much of this is because IT culture (including Microsoft's itself) in the majority of organizations has not shifted to unlock what social software in the workplace can do.

As organizations everywhere start looking for ways to weather the current economic downturn, the need to get more out of these IT investments has taken on a new urgency. As an aspect of this, it is time to take a hard look at the growing number of large, complex IT programs that failed, are failing, or are only alive because of massive infusions of over-budget cash. Just a few days ago we were talking to an executive at a large insurance company who said that their comprehensive software upgrade was in deep trouble. Our primary area of expertise is the federal government, which has had multiple failures of enterprise-wide software platforms, and continues to fund programs that may not ever deliver on their early promises. When these big programs are finally delivered—in some form or another—they are often obsolete because software has evolved one or two generations past the technology frozen into the solution.

Our contention is that the model used in the past decades for IT solution development is a "dead man walking." Things are still in transition so we cannot fully describe the future state, but we have some major indicators now and will speak to those in this chapter. Much has to do with a profound revolution in the way these enterprise social software solutions are architected. However, it is only when the new technical architecture is combined with a fundamentally different approach to IT adoption that benefits of a new comprehensive approach begin to emerge.

How do you know this new model when you see it? One government client, with an installed Microsoft SharePoint base, created and deployed an organization-wide correspondence tracking system. Elapsed time between the first requirements discussion and prototype delivery was about a week. The whole adoption strategy (change management, training, and governance) took about another week. Total cost was about $10,000. It moved so fast that it outpaced the ability of the leadership to adapt policy to accommodate the solution, which took another 6 weeks or so. In another case, a major government agency wanted to run a billion-dollar-plus competitive acquisition on a collaborative platform—dispensing with paper. The solution was delivered in a couple of months at a total cost (over and above the cost of the installed SharePoint base) of $25,000, including training. The system was used successfully and is being adapted for other major procurement efforts.

Much of what we discuss in this chapter is based on our 8-year partnership designing, building, and implementing two enterprise-scale collaboration and information-sharing programs for the federal government. One program now has over 42,000 users and includes members from government at the federal, state, and local levels; labor organizations; professional associations; and industry. The other has about 20,000 users and is federal-government-only at this point. Within these omnibus programs, hundreds of workgroup-level initiatives are developed every year, and usage statistics remain at benchmark levels for programs at this scale. To the point of cost management, the 42,000 user network is supported by 10 consultants, the 20,000 user platform by 4 full-time equivalent (FTE) consultants. Anticipating the question of whether this approach can be taken in the private and not-for-profit sectors, the answer is an emphatic "yes." Pfizer Pharmaceuticals has done much of what we describe here.

A caveat is that even though these social software or collaboration platforms can do far more than they are being used for now, they are by no means the answer to all enterprise solution problems. They are not the best native solution for massive data processing (although they can provide a user-friendly front end), and they are not optimal for highly customized enterprise resource planning (ERP) requirements (although these can be extremely expensive and often fall far short of their promises).

We would like to be able to provide you with a detailed description of the clients we worked with and a nuanced look at the hundreds of employee initiatives within the broader program framework, but we are limited by federal restrictions on disclosure and the bureaucracy involved in getting official permission to discuss client details. We will try to work around this as best we can, giving you enough context to illustrate the points we are trying to make.

Caveats aside, we can say that the two organizations we use for our case illustrations are major U.S. government departments, one with about 30,000 employees, and the other with about 20,000 personnel. Both organizations have heavy staff distribution across the United States and some people located in international offices across the globe. The bulk of their services to the public are delivered in the field. The larger organization (we call it Organization A) relies heavily on technology to get its work done, the other (Organization B) less so, although it is undergoing a major transformation effort to automate its work processes and update its underlying data layer.

Both organizations work extensively with state and local governments, and both have union and nonunion staffers. Organization A works heavily with the private sector, professional associations, and labor unions. Organization B is primarily organized around law and federal policy and relies on both for its day-to-day service to the public. Because of its regulatory role, it touches hundreds of thousands of U.S. businesses on a frequent basis, and hence the need for more IT-based automation and knowledge sharing.

Let us take each organization and talk about the challenges it faces currently. This should provide some useful background for the large knowledge-sharing programs we have undertaken and should help us make the case for an alternative model to the large, multiyear efforts we see struggling in so many places.

The first organization provides a critical safety function to the U.S. public and plays a global leadership role as well. It works with complex monitoring and technical networks internationally, and it is in the midst of a system-wide modernization. It has large numbers of its personnel distributed across the country and many overseas. It is heavily involved in regulation setting, rewriting safety procedures, training its personnel with specialized skills, interacting heavily with the Department of Defense, working in active partnership with industry and manufacturing, negotiating with its labor unions, and collaborating with private associations in the same domain. It is constantly constructing new facilities. It publishes information for public stakeholders in a dynamic and rapidly changing environment.

Prior to the beginning of our knowledge-sharing project in 2003, there was no standard, flexible collaborative platform available to offices and employees of this organization. Like many other organizations, knowledge was locked up in file-shares, obfuscated by cascading file folders, and rendered inaccessible by file names that often bore no connection to the file content. Much of the business of the organization was conducted by e-mail, which meant that organizational knowledge was further fractured by being attached to e-mails, many of which were redundant or part of inordinately long e-mail chains. In many respects, this was greenfield effort. Other things had been tried (eRoom for instance) and failed.

The second organization has about 22,000 employees, many of whom are distributed around the United States and some in foreign countries. It is highly involved in adjudicatory and regulatory processes for U.S. and foreign citizens and has a body of policy and regulation that is fractured, outdated, and interpreted in many different ways depending on the service location. It is subject to intense interest by Congress and is also going through an enterprise "transformation" process. The transformation effort is in many ways a living case study of the large, multiyear project we argue is increasingly outdated. The challenge it posed to us is that this effort is consuming the vast majority of the organization's technical resources, making it necessary for us to work on a very tight budget.

This project started because the organization was mandated by Congress (and its Inspector General) to launch an organization-wide collaboration platform. Federal employees needed to collaborate, policies needed to be revised and updated, information needed to move faster, and the babel of conflicting spreadsheets, Access databases, and file share repositories needed to be addressed. Additionally, the organization wanted to get away from an expensive and antiquated document management system and migrate into SharePoint. We inherited a nascent SharePoint program that had about 300 users and very little usage. It was basically moribund.

Our knowledge-sharing approach was similar in both organizations. In 2003, our methodology was still experimental but was highly successful over time (more on our success factors later). For the second, slightly smaller organization, we applied what we learned in the earlier project and found that we could compress the timeline experienced in the first project, achieving a much steeper growth and adoption curve. For the purposes of this paper, we describe the core approach and make differentiations between the two projects if appropriate.

We had some assumptions going in. Too many knowledge management programs collapse under the combined weight of complexity and expense (people and dollars). We had a bias against centrally controlled and highly moderated collaboration and knowledge management. We had rarely if ever seen them work at an enterprise scale. Where they did exist, they underdelivered and were extremely expensive. We felt strongly that governance and control needed to be in the workgroup, where the work happened. We recognized that in any enterprise-level system using technology there were some central functions such as operations and maintenance (O&M, security, system upgrades, etc.), but we wanted to see the business end of the system shaped and controlled by users, working within a commonly defined governance framework. We accepted early on that as this system grew, we would start to lose sight of what was happening at a granular level, and we decided that we were prepared for that. As long as governance was understood and enforced, and as long as we had ways to monitor potentially grievous problems (like the leak of personal identification information, or PII), we wanted to trust end users and put mechanisms in place to help the system police itself.

We also knew that without strong distributed leadership, the system would simply collapse as it grew. Cost considerations led us to the realization that we needed trained network administrators embedded in the workgroups (the larger client has about 1,600 of them now). These individuals would understand the workgroup's needs, configure the local SharePoint site to deliver value, and network with counterparts across the organization. They would also obviate the need for dozens of expensive full-time consultants. Finally, they would help develop and enforce governance.

This notion of distributed leadership and expertise was, and remains, critical to the management of and flow of information across both large organizations, and ultimately in the larger organizational community of external stakeholders. No one understood information needs more than embedded workgroup members. No one knew how to organize the information better than the people who had to use it. No one could devise taxonomies, keep knowledge current, and serve as knowledge brokers better than this group. We say more about what we need to help these individuals and how we embedded knowledge-sharing best practices into the structure of the network later in the chapter.

To return to the second major theme of this chapter, the collaboration/ knowledge sharing and collaborative principles and structures we just described have major implications for the design and delivery of major software-based systems in organizations. When fully and successfully implemented, these systems have thousands of users, hundreds (if not thousands) of advanced users who can configure the technical environment at the local level, and a high rate of innovation happening at all levels of the organization. The cycle between requirement and fully implemented solution can be reduced to days. Use rates, the numbers of users active on the system daily getting work done, can range between 40% and 50%.

Once the organization has a pervasive ability to adapt its information technology environment to the work it is doing, higher-order functions start to emerge. In one client organization, the procurement program decided to create an all-digital competitive bidding platform, managing all the incoming and outgoing information involving bidders on a $1 billion plus program. This capability was designed and deployed with very little consultant help for an incremental cost (over and above the costs of the larger platform) of about $25,000. Because it was created with out-of-the-box SharePoint components, it could be (and has been) modified for other competitive acquisitions, further reducing the per-application cost.

Ubiquitous collaborative literacy has another pervasive implication. The fundamentals of the collaborative system are understood by the users of the system. By "system" we mean everything from the configuration of the technology, to the method for translating business requirements into a working solution, to the governance necessary for maintaining the collaborative environment over time, and to an experience-based understanding of the promise and limits of knowledge sharing. For example, because best practices pertaining to information display, document naming, search optimization, and information "hygiene" (avoidance of multiple copies for instance) are fully integrated into the collaborative environment, odds of the environment being choked to obsolescence by outdated and badly organized information are reduced. When all this is taken together, it adds up to sustainability within a reasonable budget.

By contrast, the conventional alternative has almost risen to the level of a stereotype. There were months of requirement gathering attended by the wrong members of the workforce; months of expensive requirements analysis; requirements validation in poorly attended (and expensive) focus groups; and more months of development (often custom coded). Then a long-odds attempt to get the workforce to use the solution, which has codified outdated work processes using aging technology. In the event the system is adopted by a portion of the workforce, O&M costs start to rise because it is a custom-built environment totally dependent on a significant contractor workforce. Migration to updated software is terrifically expensive because software has to be rewritten. Billions of dollars have been poured down this drain, largely because of increasingly complex requirements and the lack of a flexible/modular approach to solution development.

We should hasten to say that we are not arguing that this is an approach that is going to fit every collaboration or technical program needed by large organizations. There are always going to be requirements for large, data-intensive systems with rigid rules that will require a more conventional "waterfall" approach to development. Our assertion is merely that this conventional approach is overused, and alternatives being made by technologies such as SharePoint are not considered because they fly in the face of technical orthodoxy.

As we move into a more detailed description of our approach, we set the scene by describing our clients in general terms and by listing the objectives and constraints we considered when moving into the engagement. We also touch on the broader implications for organizations using outmoded program models to develop solutions that need to be highly flexible, and sustainable within increasingly stringent cost limits.

What were the givens going into our first engagement? It started as a research and development effort, and it was done in the near absence of any explicit requirement for an enterprise-wide collaborative system. In 2003, collaborative software was still in its infancy, although there had been an early decision to center the effort on Microsoft SharePoint. SharePoint was brand new at Microsoft, and the company had a surprise hit on their hands but no coherent business strategy. Nonetheless, it showed promise and the effort focused on SharePoint. Although there are those who say that knowledge management (KM) or collaboration programs should be technically agnostic, we do not agree. Yes, there are inherent collaboration and KM practices that do transcend technology, but the art of the possible is shaped by technology. For instance, there are mechanisms in any technical platform that control access. These then shape governance and the roles/responsibilities of people in the environment. Facebook's access requirements are different than SharePoint's.

The whole project got started on a shoestring budget, but it had a key critical ingredient—a full-time federal government employee who had the rare combination of seniority and a willingness to behave like an entrepreneur. In the early days of the effort, he did everything from evangelism to building servers. Doing things on a shoestring budget led to another series of important decisions. He knew he would have to build the program using existing resources, so he enlisted the help of other federal employees. He knew he could not tackle really big mission-related problems, so he started solving things one small tactical problem at a time for his internal clients. He did not have money for training, so he became the trainer, and then decided to rely on a population of internal enthusiasts to extend his reach. These all became building blocks of the program and ultimately enabled it to move to scale without creating a huge expense for the organization. They remain core elements of our methodology and, we submit, set our approach apart from that of many others. Finally, they created the basic elements of what we suggest is a different approach to many major technology efforts in government.

One of the first things that happened was the identification of a governing body made up of business representatives and technical personnel. This was quite small and informal when it started, but it took early ownership of the program. As the program matured, this proved to be a critical element.

An early task of this governing body was to name the project something other than "The SharePoint Program" or "The KM Program." We needed to get the branding of the program away from the technology, per se, and focused on a mission-related phrase that would be valued by the client culture. We chose a simple three-word phrase with an acronym that did not create its own problems. This led quickly to a simple logo, and templates for the still-primitive SharePoint environment.

The solution needed to establish credibility early—that is to say that it needed to show value and relevance to people who were looking for ways to collaborate and share information. Early adopters were located and talked into experimenting with the developing program. The early project team starting looking for what they called "points of pain"—tactical problems that could be solved by the program. Because of the nature of

SharePoint technology, prototype solutions to these points of pain were developed and delivered in days (sometimes hours). This turnaround was mind-bending for internal clients and became another basic building block of our methodology.

There is a myth that some of the newer collaborative programs are so intuitive that they require no training. It is most often propagated by people with strong technical skills, and unfortunately SharePoint has suffered as a result. Even though it is much easier to use than earlier collaborative platforms, it is still not a technology that the vast majority of users can jump into with confidence without training. Moreover, the argument that computer-based training (CBT) modules are sufficient was also proven wrong. Most of the clients in the organizations we work in do not have the time or the foundational knowledge needed to move efficiently through a self-service CBT program. Also, most CBT programs only teach what we call "buttonology"—they do *not* put SharePoint in the context of the organization, and they do not teach best practices. For example, we discourage nested folders because they can be a subterranean sink for valuable information. We also teach admins how to manage access and security in a federal government setting.

Training became and remains a critical part of our approach to collaboration. As our engagement with this first major client unfolded, the training program eventually became a basic two-tier curriculum. Our 101, or basic user training, included modules on knowledge management, features of SharePoint relevant to what they needed, best practices, and basic security. Our 201 training was meant for super-users or admins (we called them "facilitators") and was designed to equip them for their roles.

If you recall from our earlier point, reliance on people embedded in workgroups was in part necessitated by resource constraints. It also made inherent good sense for growing and sustaining a knowledge-sharing/collaborative environment for reasons also discussed above. In the case of this first client, it was where we experimented with and developed the methodology for creating and sustaining a community of facilitators which spanned the organization. We created and polished our 201 training curriculum. We instituted a regular conference call where facilitators could get help from us and each other—it actually became a highly productive community of practice. This organization now has something on the order of 1,600 facilitators, many of whom can do SharePoint configuration at the local level using out-of-the-box Web parts. They know where to come for second- and third-tier help if they need it. In many instances, they are driving the creative edge of the collaborative environment. We often said that we are at the center of a rapidly expanding collaborative universe (which has basically grown exponentially) and although we cannot possibly know what is happening in the thousands of SharePoint sites in the system, we do hear of some from time to time. One example was the linkage of construction teams who were all working on similar projects across the country. Once they discovered that they could share files, they began to exchange computer-aided design (CAD)/computer-aided manufacturing (CAM) drawings, lessons learned, best practices, and program management tips. We learned of this well after the fact. The critical point? *They did not need to come back to a central IT office somewhere to get permission to do this, or hire expensive consultants to create the capability.*

The final point on this first program is that we developed a good method for measuring how things were going. Because he was a research and development (R&D) guy at heart, the founder of the program found a metrics tool that would measure system activity all the way from the individual to the enterprise. He actively used these measures to determine his growth leaders, which areas had ceased activity, and where things were showing early signs of stagnating. He also learned how to identify signs of health in the system (distribution of people using the system, level of activity, etc.) and how to make decisions about where to intervene. For instance, SharePoint sites have their own life cycle. Some lived out their period of utility and need to be closed and archived. Others are expanding rapidly and, if not attended to, could exceed storage limits on the system. Still others are really only used once a year or so for annual reviews and episodic use is OK. The main point was that there were ways to gauge the health of the system. It also helped the governing body show relevance to senior leadership and people who controlled the organization's budget. Let's summarize some of the key aspects of this first program and then jump ahead in time to update you on where things are today, including some of the mid-to-late life cycle challenges the program is facing. Here are some of the key features of the approach:

1. Governance by business owners
2. A dedicated full-time director or leader
3. Tactical approach to problem solving
4. Rapid turnaround of solution prototypes
5. Facilitator-led, customized training
6. Facilitators or trained admins embedded in workgroups
7. Locally led innovation and sustainment

 8. Governance enforced by facilitators and a representative governing body
 9. Supportive, not controlling role by the IT shop
10. Metrics

Let's fast-forward to the present on this particular program. It has somewhere on the order of 42,000 users, several thousand SharePoint sites, and 1,600 facilitators. Training courses remain booked to capacity in the program's seventh year. It has become a collaborative/KM commodity. It is supported by a small core consulting support team (about eight people), with a help desk of three (for 42,000 users). On the scale of federal government expenditures for IT programs, the budget is extremely modest relative to the overall impact of the program. There are several terabytes of data on the system. It recently went through a successful migration from SharePoint 2003 to 2007, and is looking to SharePoint 2010.

As the program has matured, business ownership has shifted away from the core program team to the actual workgroups and organizational divisions within the larger agency. We would argue that this is not necessarily a bad thing, and that it might be an aspect of the life cycle of an effort like this one. We do think that increased business ownership at a senior organizational level would increase the cross-agency impact of the program—accelerating the movement of applications and lessons-learned from one division to another. It would also counter the tendency of these kinds of programs to drift back into the control of technologists and conventional "information technology" management approaches. Fortunately, the cumulative effect of the program created a realization among the workforce that the means for collaboration are available and now underpin a lot of what they do. They have come to take it for granted, which is a good thing. Collaborative literacy has become nearly ubiquitous, even if it is uneven (and will likely remain so). What we have not yet been able to do, and would like to at some point, is take a rigorous measure of how the program has affected productivity over time. The major challenge is that enough years have gone by that we lack data on the "before" state.

Finally, we think this program offers a very rich research environment for scholars and practitioners looking to examine the nature of large-scale collaboration that developed a measure of maturity. Have collaborative and knowledge-sharing activities started to transcend the workgroup? Are workgroups now starting to collaborate on larger-order problems or is the vast majority of the activity still centered on the smaller groups?

Let's move to the second client. Much of the story here centers on how we took what we learned in the first program and applied it to a fresh environment. Most of what we focus on is where and how things differed to show how the methodology and technology evolved.

When we started the second program, we were working with technology that had changed radically since our early days on the first project. Microsoft invested a great deal in SharePoint, clarifying earlier confusion between collaborative and portal technology, improving the user interface, and refining features like automated workflow, spreadsheet integration, and the development of custom data entry forms (or "lists" as they are called in SharePoint). This meant that facilitators in workgroups had a more powerful set of tools at their fingertips, and because the software was easier to use, we got deeper support from facilitators more quickly.

We brought several years of experience and learning at the first client to the second. There was little or no trial and error getting started. We moved very quickly to set up a Strategic Planning Team, name the program something that had mission relevance, and find some early adopters who were interested in getting started quickly. We also had an experienced trainer and help desk person, and they were able to get right to work. We were joined by a very strong Microsoft technical consultant who had technical and people skills and was able to start producing solution prototypes almost immediately. He was also able to connect to other Microsoft personnel in the broader organization and to Microsoft's many technical resources. He eventually became a powerful advocate for the adoption model, which is rare in the world of tech-centric service providers.

There was a thirst for collaboration in this client system driven both by business need and a greater awareness of collaborative technology by the workforce. Once we started to produce quick prototypes that solved problems, the word spread quickly and the program expanded through demand-pull instead of supply-push. As an example, when we met with the director's office in the organization, their "point of pain" was executive calendar management, the need to merge and coordinate multiple calendars, and to incorporate input from other offices on things like travel logistics, security, and local liaison. We knew from experience that a quick response on a "point of pain" identified by the client created a starting point for increasingly complex solutions.

Finally, we should say something about the role of "executive director." In both engagements we had an executive director who actually helped shape the methodology. He understood where controls needed to be tight (overall information security, standardized templates, management of PII data, etc.) and where the reins needed to be loose (local configuration, for example). He was also skilled at finding the resources needed for the program and knew how to work with a small consulting team.

We started with 300 users and probably a tenth of those were actively using the collaborative system. Within a year we were at about 16,000 users, with a use-rate of about 45%, which meant that roughly 7,200 users were on the system every day doing work. We just got word that the chief financial officer (CFO) is using the system to do a critical organizational data-call on performance achievements, which means that the system is being used at the enterprise level for core business functions. The training courses are overbooked, and our more experienced facilitators are starting to serve as internal consultants and trainers, which allows us to support now close to 20,000 users with a team of about four full-time equivalents. To give you an idea of costs, we helped one office create a comprehensive document clearance platform in less than a month (most of the time was spent clarifying the clearance process, not building the SharePoint site). This meant that they had a fully delivered system for under $10,000. The client—someone with decades of federal service—said it was the fastest technical development project he had ever seen.

As we proceed with this client, we are starting to see interest in multioffice solutions, like records management, policy coordination, and publishing, and advanced knowledge management capabilities. This is the kind of progression we look for and support, and it works because it is growth at the client's pace, not the pace of the technology. There are hundreds of tech-centric programs that fail because technical teams are trying to move the client at their pace. Talented technical people often do not have the patience to start with mundane projects like calendar management. We knew from experience that as client personnel gained confidence in the technical environment, they would drive solutions of increasing sophistication. The risk of implementation drops radically because we have trained clients who are designing their own solutions at their own pace.

Allow us to switch back to our second major theme at this point. This second effort went from near zero penetration to near saturation (all offices in the organization, and 85% of the entire workforce registered on the system) in a little over a year. The level of collaborative literacy is far higher than it was a year ago. Solutions are being developed in days at a tiny fraction of what it costs in more traditional models. We are now starting to see interest in higher-order enterprise business functions like performance management, work-flow automation, and records management. With a base of trained users, and trained facilitators, the elements are in place to design and implement these programs, because they will be driven by users, substantially configured by users, and trained by users. The development process is often so fast that traditional "Agile" mechanisms (scrums, sprints, etc.) often slow things down. The software continues to improve with the release of SharePoint 2010 and the interrelationship of many products (both Microsoft and non-Microsoft) with the SharePoint platform.

Ironically, there is a very large traditional program being run in this organization that is way behind schedule and over budget. It is already being scrutinized by Congress and in our view is likely to fail because of its complexity and cost. To give you an idea of scale, our total program is about two tenths of 1% of the other program. Could we do much of what they are doing? Probably. Are the politics and technical culture prepared to accept our approach as an alternative? Not yet.

Here again, this represents a great research opportunity for a student of collaboration and knowledge management. How do you measure the cumulative effect of a system like this? We have isolated anecdotes (one process got reduced from 5 hours to 20 minutes), but how do you assess the value in a system that is not profit driven? How do you describe the nature of knowledge sharing in a system like this? It does not happen in neatly defined knowledge repositories, there are no formally organized pan-organizational taxonomies for explicit knowledge. In this system, knowledge is being shared in thousands of different ways every day. It is communicated in announcements, conveyed in collaborative spreadsheets, entered into custom databases, produced in collaboratively authored documents, and exchanged between headquarters and field elements in a steady stream. How does the KM or collaborative theory base apply to this kind of program?

Where do we go from here? Chapters like this help us take time to codify what we learned and the issues we see. We are already embarking on a third major project of this type (larger in scale and more complex), and we expect our basic methodology to continue to evolve. We developed a highly productive working partnership with Microsoft, which is beginning to understand the power of marrying a powerful adoption methodology to software like SharePoint. We hope to see that continue to evolve and to continue our exploration of a joint approach. We hope to refine our approach to metrics, perhaps embedding performance sensors in our collaborative environments to give us a good sense of what is happening at the intersection of people, work teams, and technology. We would like to be able to have the system flag a failing collaborative effort in time to allow us to choose and deliver an intervention strategy. We look forward to incorporating social media more effectively into our knowledge-sharing efforts.

To conclude, we are convinced that we are on the cusp of an era that better merges people, organizations, and technology. We think that technology has outpaced the people and organizational aspects of integrating all three elements, but we think our approach contributes to a new model. We are realistic enough to know that today's IT culture is going to take time to change, but we are absolutely convinced that the software companies willing to embed the kind of approach we advocate will leap ahead of their competition and create major value in a world of smaller budgets. We look forward to the challenge.

14

TATA Chemicals—Knowledge Management Case Study

B. Sudhakar and Devsen Kruthiventi

ABOUT TATA CHEMICALS LIMITED (TCL): SERVING SOCIETY THROUGH SCIENCE

Tata Chemicals Limited (TCL) is a global company with interests in businesses that focus on living, industry, and farm essentials (LIFE). The story of the company is about harnessing the fruits of science for goals that go beyond business. This story began in Mithapur, Gujarat, in western India with the creation of a plant that would raise a wealth of marine chemicals from the ocean, with the potential to touch human lives in many ways. From these humble beginnings an international business in industrial essentials was created, with operations across four continents. Tata Chemicals is the world's second largest producer of soda ash with manufacturing facilities in Asia, Europe, Africa, and North America. The company's *industry essentials* product range provides key ingredients to some of the world's largest manufacturers of glass, detergents, and other industrial products.

Through its *living essentials* portfolio the company has positively impacted the lives of millions of Indians by using Salt as a carrier of Iodine. Tata Chemicals is the pioneer and market leader in India's branded iodized salt segment. With the introduction of an innovative, low-cost, nanotechnology-based water purifier, it is providing affordable, safe drinking water to the masses, thereby preventing people from suffering from waterborne diseases. With its *farm essentials* portfolio the company carved a niche in India as a crop nutrients provider. It is a leading manufacturer of urea and phosphatic fertilizers and, through its subsidiary, Rallis, has a strong position in the crop protection business. The Tata Chemicals Innovation Center is home to world-class research and development (R&D) capabilities in the emerging areas of nanotechnology and biotechnology. The company's Center for Agri-Solutions and Technology provides advice on farming solutions and crop nutrition practices. The company also entered into a joint venture (JV) with Singapore's Temasek Life Sciences Laboratory (Joil) to develop jatropha seedlings to enable biofuels capability. In line with its mission, "serving society through science," the company is applying its expertise in sciences to develop high-tech and sustainable products.

Science for Sustainability

Tata Chemicals' mission, vision, and values are deeply rooted in the principles of sustainability. For the company, sustainability encompasses stakeholder engagement, environmental stewardship, creating economic value, promoting human rights, and building social capital. Tata Chemicals supports the UN Global Compact and is committed to reporting its sustainability performance in accordance with Global Reporting Initiative (GRI) guidelines. The company actively works toward improving its eco-footprint with a policy of "avoid, reduce, and reuse." Resource optimization, alternative sources of fuel and raw materials, and maximizing reuse and recycling are key drivers in operations. The company was recognized for its clear commitments to sustainability and its good environmental management practices.

Community Services

In 1980, Tata Chemicals set up a nongovernmental organization—Tata Chemicals Society for Rural Development (TCSRD)—that works toward holistic community development, including managing water, land, and other natural resources; encouraging enterprise development; and promoting health and education. TCSRD's activities were recognized at a national level. Tata Chemicals Europe (formerly Brunner Mond) is a major sponsor of the Lion Salt Works Trust, a local heritage project in Cheshire, United Kingdom, and of the Weaver Valley Initiative, part of the path-breaking Mersey River cleanup campaign. In Kenya, Tata Chemicals Magadi supports local healthcare facilities and works to provide education, water, and employment opportunities.

Tata Chemicals is also involved in efforts to preserve the biodiversity of land along the Gujarat coastline and the nesting sites of migratory birds. TCL and Wildlife Trust of India (WTI) signed a Memorandum of Understanding (MoU) for a conservation project that will create awareness and undertake research to save the endangered species of whale shark that visits the coastal shores of Gujarat.

OVERVIEW OF KNOWLEDGE MANAGEMENT (KM) AT TCL

TCL always had a strong focus on knowledge since it indigenously developed soda ash manufacturing know-how and was the first to produce it in India. However, post-2000 TCL transformed from being a one-location manufacturing plant (Mithapur, mainly salt and soda ash) to multisite, multigeographic, multiproducts (soda ash, bicarb, salt, a range of agricultural inputs like fertilizers/crop protection chemicals, water purifiers, sulfuric acid, and many more). TCL has a strong performance orientation, great focus on its customers, and a culture of excellence in everything that it does. Tata Group adopted an excellence model, Tata Business Excellence Model (TBEM), based on the famous Malcolm Baldrige Model that has knowledge management as a key area of excellence. TCL was assessed and graded well by TBEM assessors consistently over the past years. TCL, being a learning organization, reinvented itself over the past few years to meet various business challenges. Because there is great diversity in terms of geographies, products, and sites, the need for structured knowledge management was felt and in 2003 a KM vision and purpose were created along with KM strategy (Figure 14.1). The main focus

<u>Vision</u>

• To create a culture of systematically harvesting and sharing knowledge in TCL, in a manner that employees are encouraged to continuously improve their own knowledge and the skills that would be required to manage both the current and future needs of the business.

<u>Purpose</u>

• Knowledge Management will result in greater functional excellence, qualitatively better decision making, reduce duplication of work and free up time for key business imperatives.

FIGURE 14.1
KM vision and purpose.

of KM is on the individual to inculcate the culture of sharing and seeking. This is incorporated in the company's core values as "Caring." TCL believes people are the fulcrum of KM and to bring about a change in their behaviors toward more effective collaboration is the key to success in implementing KM.

> As we progress deeper into the information age, knowledge will be the key differentiator, and embedding knowledge management in our company is critical for its progress.
>
> **R. Mukundan MD, TCL**

KM Strategies, Processes, and Culture

KM Strategy

TCL implemented the balanced scorecard across all its businesses. The KM strategy at TCL is designed in such a manner that it is aligned with the business objectives of TCL. The strategy of KM at TCL has a LTSP (Long-Term Strategic Plan) that is aligned to the LTSP of TCL. Knowledge in TCL is classified into three levels:

Level 1: *Basic needs*: Delivering basic knowledge needed by the organization to improve operations or solve problems.

Level 2: *Enabling needs*: Delivering knowledge needed to achieve its tactical goals, collecting key business intelligence, promoting reuse, making connections with experts and obtaining information as needed, facilitating greater integration, and identifying information or knowledge problems related to business processes.

Level 3: *Strategic needs*: Providing proactive support in guiding senior management to make use of knowledge for management support, decision making, and innovation. KM strategy makes sure that all the needs of the business are sufficiently met.

Key improvements of KM in the last 6 years are shown in Figures 14.2 and 14.3. (Figure 14.2 depicts the knowledge flows in the company.)

KM Processes

All the business and support processes at TCL follow the Enterprise Process Model (EPM) methodology consisting of three levels of deployment with a process owner for each of the defined processes. KM processes

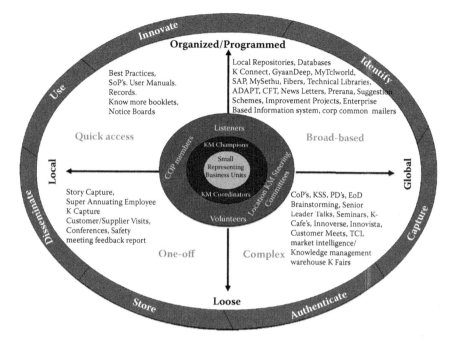

FIGURE 14.2
Knowledge map.

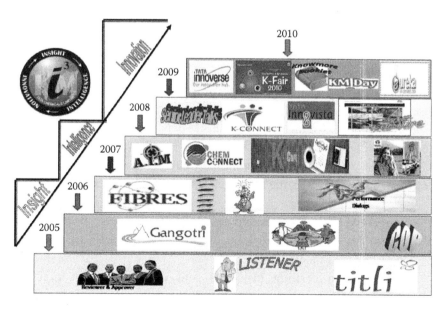

FIGURE 14.3
Journey from insight to innovation.

as per the strategy of the company are a part of continual improvement processes. There are three distinct EPM level III KM processes:

- Process for Knowledge Identification, Capture, and Review
- Process for Knowledge Sharing, Replication, and Usage
- Process for Idea Management and Implementation

Each of the processes are well defined with a set of efficiency and effectiveness measures which are tracked and reported regularly.

There are many KM initiatives at TCL to address KM processes in a systematic and innovative manner to bring out best practices that help in meeting the business needs and challenges. There are various KM initiatives that target people at different roles/levels in the organization. All the initiatives are regularly monitored, and a process of evaluation and improvement (E&I) methodology is adhered. All the KM EPM processes undergo periodic DEMMI (Define, Measure, Manage, Improve) and are thus constantly improved. Further, a benchmarking study of KM processes is undertaken with other Tata Group companies and also identified relevant companies from across the globe. This procedure is called BEEP (BEnchmarking Every Process). These processes derive strength from the culture of continuous improvement consistently demonstrated by employees.

The core KM team of TCL is composed of a small dedicated group of knowledge managers representing core business units across all business units and support functions. This team has the ability to understand the unique business needs of the Strategic Business Unit (SBU) and support functions analyze the needs and cater accordingly. In addition to the core group, each department of TCL has an identified KM coordinator to function as one point of contact between the KM department and his or her own department. Further, there are a set of listeners who function as extended KM department members on a voluntary basis (Figure 14.4). Each of the knowledge managers has the following roles and responsibilities:

- Spreading the awareness and ensuring effective functioning of KM initiatives within their own business units in line with specific business needs of the BU
- Accelerating learning processes and identifying best practices through KM practices for improved efficiency
- Developing KM processes, procedures, and applications and implementing these in their own work environment

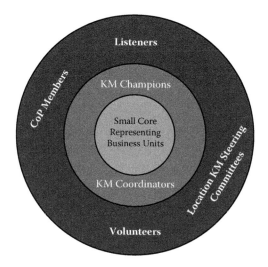

FIGURE 14.4
Employee involvement inclusive approach.

- Promoting knowledge sharing culture and building collaborative mechanisms to capture and encapsulate knowledge to thrive in the future
- Ensuring KM initiatives adherence to plan
- Tracking department-level KM activities
- Facilitating department-level and site-level KM activities
- Reporting KM data on a monthly basis
- Conducting monthly review of department KM coordinators and listeners of the business
- K Connect (KM Portal) management for the business unit
- Coordinating with other business units on KM activities
- Facilitating faster knowledge transfer among employees (specifically targeting super-annuating experts, new joinees, and other senior employees)

KM Culture

With a view to create KM culture in the organization, the company invested significant management time to till the soil and sow the seeds. Many interventions were undertaken to create critical mass of employees who can lead the transformation of creating KM culture, like programs on extension motive, quizzes, theater shows, skits, listener programs, exposure visits, benchmarking visits, and so on. The senior leaders used the existing communication forums to talk about KM culture.

Today KM at TCL is viewed as a major contributor to building intellectual capital in the organization encompassing free flow of knowledge across all geographies it operates in. The intent is to make KM a way of life at TCL and focus on conscious learning through continuous realization of the need to build the knowledge base of the organization. As mentioned earlier the main focus is on the individual to inculcate the habit of seeking and sharing knowledge. The basis for KM initiative and process design is to consciously identify what knowledge is critical to the organization, what knowledge already exists, what still needs to be known, where to get this knowledge from, and how efficiently and effectively this knowledge is put to use to meet organizational goals and challenges. The initiatives reinforce conscious learning and improvement philosophy through a collaborative approach to knowledge sharing. The reward and recognition around knowledge management activities are structured in a manner to strengthen and nurture the KM culture of continuous improvement, collaboration, knowledge seeking, and sharing.

Key KM Initiatives

- *Titli—Story Capture* TCL is more than 65 years old, and with this vast experience, what comes naturally is a huge pool of tacit knowledge among its existing employees, with experience ranging from a few years to almost 35 to 40 years of hands on REAL anubhav. All these knowledge sources are available for carefully and sensitively "harvesting," documenting, codifying, and making knowledge available for "sharing" and "seeking" to employees across the locations. Titli (a butterfly) is the name for this initiative of capture of tacit knowledge and conversion to useful K-nuggets. A butterfly stands for many colors and helps in cross-pollination. Our Titli also is intended for cross-pollination of ideas, experiences, and learning from each other's failures and successes. It has varied hues of each individual's perceptions of instances in the form of ANUBAV or stories as its core inputs. As a part of this Titli, TCL developed and implemented a unique concept of listeners. Those employees who volunteer to be a listener are provided structured training into active and passive listening. A listener is also trained to capture tacit knowledge in the specified formats and submit to the KM portal or hand it over to the K coordinator of that department. Stories elaborating experiences of employees and their learnings and opinions were shared since the beginning. Stories are rated on their impact and sustenance of impact, based on whether they led to process improvements, savings accrual, addition of capabilities, changes in Standard Operating Procedures [SOPs], etc. There is a clear defined process of refinement and distillation of these stories. Accepted stories are being converted into case studies/use cases to create a database of distilled stories, best practices, and explicit documents that may be shared cross location, thereby facilitating the best practice sharing and learning. The approved stories with high ratings for ease of replicability, applicability to TCL, and learning value are converted into case studies/use cases and are published quarterly, thus helping the end user of knowledge to quickly identify cases that can be replicated in their respective departments (Figure 14.5). Case studies/used cases mention the area of impact of the improvement made, ensuring effective utilization of the knowledge case.

- *Community of Practice (CoP):* There are various CoPs in TCL. These sponsored CoPs have a charter and a set of deliverables. Some of the CoPs have Sub-CoPs, too. The basic charter for all CoPs has three major objectives:
  - To enable colleagues to learn from one another through the sharing of issues, ideas, and lessons learned problems and their solutions, research findings, and other relevant aspects of their mutual interest
  - To more broadly share and better leverage the learning that occurs in CoP with others outside the community

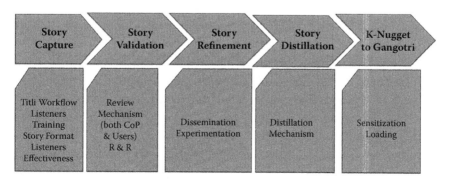

FIGURE 14.5
Converting tacit input to explicit knowledge nuggets.

- To generate tangible, measurable, value-added benefits to the business
- *Know More Booklets:* Each department comes up with a simple booklet addressing a particular area of its functioning, wherein the present practices, the best practice, and the learning that stimulated improvements are incorporated into the booklets. These are regularly published along with the help of the KM department.
- *Senior Leader Talks:* Every month there are talks given by a senior leader from the executive committee to all TCL employees via webex and video conferencing, wherein valuable learning and knowledge are shared. These talks provide a platform for knowledge dissemination from the senior leaders.
- *K-Capture from Super Annuating Employees:* A major KM challenge within a large size organization is capture and transfer of knowledge from experienced staff members who are about to retire. It is a common organizational observation that when experienced employees superannuate, they create a vacuum behind them that cannot be filled despite the best documentation efforts of their activities and responsibilities. This occurs because documents essentially capture only the manifest information while leaving behind a considerable mass of contextual knowledge, embedded ideas, and best practices that together form the tacit knowledge base of the organization. Because tacit knowledge is carried as individual memory, when employees leave, the overall organizational memory also suffers a setback that eventually causes reduced performance and efficiency. For capturing the tacit knowledge, based on the case a K-capturer is mapped with superannuating employee and on a one-to-one discussion, the outcome of the same are captured in the explicit form called "Memoirs." TCL has separate processes in place for capture and transfer of knowledge from each employee.
- *Knowledge Sharing Sessions:* Every department comes up with a calendar of knowledge sharing sessions that it would like to conduct every month. These provide platforms for competency building, brainstorming, identifying experts, and also getting relevant external inputs into the system.
- *KAM, Customer/Dealer, Supplier Portals:* Key account manager workshops (KAM) are conducted regularly to address not only various issues but also look at the K-transfers needed for appropriate customer focus and intimacy. There is a CRM and SCM portal

that captures customer needs, and these are then provided to the concerned departments for action and capability building. KM plays a vital role in the latter in terms of identifying and providing the relevant experts and the required knowledge base for the same.

• *K Connect:* Given the company's growth rate both in terms of knowledge and people, several collaborative solutions were explored for building a KM portal. In 2007, TCL created K Connect; an interactive Web portal that helps manage knowledge assets and run virtual collaboration forums (Figure 14.6). K Connect helps in connecting people to content and people to people. With key components of K Connect in place, adoption levels of the portal solution reached new highs. These initiatives promote knowledge sharing and learning among different categories of end users and help create a connected organization promoting creativity and innovation. K Connect has various features like Post a Document (through this one can post explicit documents to build a knowledge repository for TCL. All the documents posted go through peer and expert rating before being accepted to be considered for knowledge repository), Search a document (normal and advanced search), Browsing Knowledge Pieces (like RCA's, best practices, EPM improvements, safety-related improvements, etc.), and Contribute a Story (here one can contribute tacit knowledge in terms of experience, learning, and success).

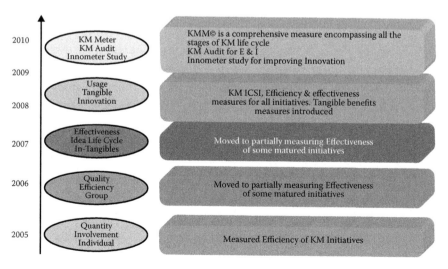

FIGURE 14.6
Refining KM measurement.

The stories contributed go through a refinement and distillation process through an automated work flow in this section, K Champs (here we can find a list of KM coordinators, listeners, award winners in various KM programs, etc.), KM Usage (in this section one can log-in KM usage, be it in terms of various parameters like source of learning, area of impact, tangible benefits accrued because of specific KM intervention, etc.), *AskTheExpert* (an expert locator that profiles list of self-registered/nominated experts across the organization who with their expertise aid in speedy resolution of problems. This tool helps locate and contact experts in a wide range of technical and functional areas seeking answers to specific queries. In TCL we identified 487 areas of concern for business, Senior Leadership Speak (includes transcripts and recordings of all senior leadership talks for reference), KM Help Desk (any question or need to be addressed by the KM department is posted here for immediate attention), *My K Connect* (a personal space for every employee to share his or her profile, interests, and areas of expertise. Each individual also gets a view of their contributions and collaboration on the Connect portal). *FIBERS: Fact and Information Based Reverse Engineering of Strategy* is a portal to capture competitor information and then use it to reverse engineer the key processes to adapt to the change. Fibers is an integral feature of K Connect. This is one of the key ways in which KM actively helps in strategy of TCL.

- *MySethu:* Because safety is a key focus area and a value in TCL, this section in K Connect provides a way to log in all learning's, best practices, etc., specific to safety. OTIS is an online engineering design document library that is available across locations for select users and has around 80,000 documents. A link to this repository is also available on K Connect. Recognizing that knowledge is the guide forcing our business, we rapidly adopted social networking to facilitate connections between people who share similar interests and pursuits. This builds an inclusive knowledge-sharing and learning culture in which employees are empowered to address their development needs; to make cross-organizational connections with one another regardless of role, department, or location; and to engage in informal learning. The following collaboration portal offers services to create, use, and share information. *MyCoP* Section on K Connect (every CoP is provided a portlet that is managed by the CoP members to interact with each other and share information within themselves. These special interest groups promote and fuel common interest among employees across technical, domain, and functional areas. It is an open forum to debate, discuss, clarify, share information, seek answers or solutions, and collaborate in any way to share knowledge).

- *Blogs:* To encourage discussion and open collaboration and brainstorming, Blogs were incorporated into K Connect. Blogs promote expression of viewpoints and ideas. Blogs are also used by a number of departments to share general information about activities being carried out in each department. *Wikis* are used by the IT department to distribute information regarding the IT environment and fixes. All users get points based on the activities he or she undertakes on K Connect. These My Privilege points can be cashed in based on the reward scheme available on K Connect and seen on the My Account section.

- *Yammer* is an enterprise social network service launched in 2011 in TCL. Unlike Twitter, which is used for broadcasting messages to the public, Yammer is used for private communication within organizations or between organizational members and predesignated groups. TCL has many active members on Yammer, and it is a powerful platform for enterprise social networking.

INNOVATION AND KM

The tag line of knowledge management @ TCL is "Insight2Intelligence2 Innovation." Innovation is a key enabler and differentiator for TCL which helps drive learning and innovation in the changing business scenarios. TCL has a world-class innovation center at Pune focusing on nanotechnologies and biotechnologies. We also have a Center for Agriculture and Technology (CAT) at Aligarh to provide advice on farming practices in general and crop nutrition practices and solutions in particular. The company entered a JV with Temasek Life Sciences Laboratory Ltd. (TLL), Singapore, for Jatropha seeding. Further, all KM practices in TCL accelerate innovation through effective management of collective ideas, insights, wisdom, and experiences, maximizing on the intellectual capital. Leveraging on the knowledge-sharing culture in the organization, TCL promotes creativity and innovation among employees to enhance organizational and personal value and develop competitive advantage. We believe in combined power of people, processes, and technology to enhance organizational performance and achieve sustainable innovation.

- *K Café:* One of the key initiatives of KM around idea generation is K Café. The K Café is an innovative yet simple methodology for hosting conversations about questions that matter, based on the needs of the business. Teams are invited to come together to present ideas. The ideas are evaluated on six criteria: innovation, novelty, actionable, interdisciplinary approach, futuristic, and impact on business. All the ideas accepted are tracked in terms of its implementation status. K Café is also a feature available on K Connect from where one can see the upcoming K Café themes, nominate for participation, and look at the conducted K Café ideas and their status.
- *Tata Innoverse:* This is a Web 2.0 application of TQMS (Tata Quality Management Services) where challenges can be posted and ideas for these challenges can be received from participating Tata Group companies. TCL became a part of the Tata Innoverse in January 2010. The link to Tata Innoverse is through K Connect.
- *Prerana:* This is a suggestion portal open for all employees in TCL so they can post a valuable suggestion for any area of concern for TCL. A link to this is also provided through K Connect.
- *Eureka On Demand:* Another feature on K Connect is *Eureka On Demand.* This is a platform for enabling a collaborating problem-solving approach across the value chain of the organizational problem-solving process—from problem definition to solution finding. As opposed to our regular or conventional problem solving, in which talent resources of any given or limited geography are used. In this collaborative approach, by connecting dispersed talent resources such as employees of other global locations, retired employees, educational institutions, research organizations, original equipment manufacturer (OEM) suppliers, vendors, and channel agencies, every employee has access to the complete ecosystem of TCL. This application takes care of the type of non-disclosure agreement (NDA) signed with our external collaborators and suitably involves the stakeholders in problem solving.
- *icare:* Tata Chemicals conceptualized and launched in 2011 a unique program called icare, with the following objectives: shift the focus for employees from mere ideation to evaluation of the concept from a holistic business perspective; provide required financial and other

support for building prototypes or market-ready products; facilitate in finding appropriate internal and external connects and support required for creating the product or service; design the marketing and distribution strategies for the product/service; and finally, provide accelerated trajectories or paths, in both financial and career terms for those who demonstrated a "make it happen" approach of converting ideas to business realities.

- *Sharing TCL Knowledge with External World:* The KM department started conducting a "K Fair" about 3 years ago. This is a gateway for TCL to share its internal knowledge with external collaborators like academic institutions (from where TCL goes for recruitment), OEM partners, vendors, and suppliers. Further, TCL regularly participates in various technical conferences, industry forums, and government regulatory committees.

- *Kisan Help Line:* The KM department developed in coordination with the TCL CAT research center, Aligarh, and TBSS (Tata Business Service Solutions) a toll-free-number-based help line for farmers in the region where TCL sells its Agri Products. Through this service any farmer can call this toll-free number and get solutions to his crop-related problems. This service also ensures primary, secondary, and tertiary assistance to the farmer for the crop-related issues.

MEASURING KM

Measuring the value of knowledge and ensuring that it is in line with the business objectives is critical. The TCL KM framework evolved to ensure the continuous feedback of results into the system (Figure 14.6). This allows TCL to continuously refocus the KM strategy, the key business processes, and infrastructure. TCL has two focus areas while measuring KM. One is focused on individual and the other is as a group. Individual KM was measured through privilege points one accrued by contributions to KM. For measuring group KM, TCL developed a unique framework to measure the effectiveness of the KM Meter (KMM©). The "Knowledge Management Meter" (KMM) was proposed as a parsimonious and useful tool to help TCL gauge its knowledge management capabilities. The first step is understanding the difference between what TCL is currently doing what it needs to do in order to maintain and improve. At the macro level, the KMM enables locations to compare themselves with each other. At the micro level, it calls the attention of the departments to areas needing improvement in current and future knowledge needs. In either case, KMM provides a robust indicator and basis for decision making and organizational support and development in terms of knowledge management. KMM is a comprehensive measure encompassing all the stages of KM life cycle. KMM is mapped to the five levels of the TQMS Knowledge Management Maturity Model. All the key effectiveness and efficiency measures of Knowledge Management EPM (enterprise process Maps) level II process are reported to the EXCOM every month along with trends. (The EXCOM is the highest body of TCL.) In addition, every business unit reviews knowledge management metrics relevant to it in its monthly review meetings.

LESSONS LEARNED DURING THE KM JOURNEY

- Leadership plays a vital role in making KM become a culture in the organization.
- Preparing the soil is very important. Developing critical mass of employees who appreciate this is important for sustainability.
- Moving KM from "nice to have" to a "way of life" requires enormous effort, constant reinvention, and patience.
- Constantly aligning KM strategy to business strategy makes it more useful and also more visible.
- A hybrid of top-down and bottom-up approaches needs to be followed to make KM work.
- Reward and recognition programs definitely play a role in promoting KM. However, they constantly need to be revisited to create fresh targets and motivation.
- Peer recognition is found to be a prime motivator.
- KM strategy should focus on people in the organization, and not processes or technology, as they are the real players.
- Technology enablement of KM helps hasten the KM adoption process. The KM portal has become a powerful and mandatory channel for eliciting both explicit and tacit knowledge.

- KM requires a balance between culture and metrics in organizations. While KM is all about culture, measurements and return on investment (ROI) are required for constant benchmarking and improvement.

In conclusion, KM is a continuous journey from insight to intelligence to innovation. And as Robert Frost said, "The woods are lovely, dark, and deep, But I have promises to keep, And miles to go before I sleep, And miles to go before I sleep"...*so is the KM journey at TCL.*

15

Knowledge-Enabled High-Performing Teams of Leaders

Bradley Hilton and Michael Prevou

The leaders who work most effectively, it seems to me, never say "I." And that's not because they have trained themselves not to say "I." They don't think "I." They think "we"; they think "team." They understand their job to be to make the team function. They accept responsibility and don't sidestep it, but "we" gets the credit. This is what creates trust, what enables you to get the task done.

Peter F. Drucker

PREFACE

A dramatic transformation is underway, changing how the world connects and collaborates. This change will ultimately have a significant impact on how we work to create, develop, and produce goods and services. Geographically dispersed, cross-boundary, and multifunctional teams are often a necessary component of today's global work environment. The diversity of these teams helps us work through the complexities of organizational hierarchy and bureaucracy, as well as connect us to the diversified expertise required to produce valuable goods and services. Traditional hierarchical structures are often being dissected and deconstructed to form these teams, necessitating whole new approaches to business and government alike. Today, connecting an organization's workforce with experts, vendors, partners, and customers is absolutely necessary but insufficient to facilitate the sharing of essential knowledge and to remain competitive. The speed with which knowledge is applied for effective decision making, creation of new products and services, political and regulatory solutions, and innovation requires a deliberate and practiced framework to make those connections productive.

In this chapter, we address knowledge management (KM) as it pertains to enabling high performance in cross-boundary and multidisciplinary teams. We discuss KM as a means to improved performance and increased team productivity. To do this, we offer a new, more comprehensive framework for managing the components of the knowledge environment, rather than trying to manage knowledge. We offer an approach that we have been working with over the past several years to deliberately form and launch teams. We discuss the conditions required for high-performing teams to exist in an organization, and we talk about case studies in which we applied and researched various techniques on diverse, sometimes geographically separated teams. Our hope for this chapter is twofold: first that you will consider the value of looking at knowledge management in a different way and second that you will acknowledge the value of a deliberate approach to forming, launching, and developing any team within your organization and, when given the opportunity, ensuring your teams are properly supported.

INTRODUCTION

Current knowledge management (KM) publications (as well as most KM projects we observed) usually begin with a discussion of the need for the balanced application of people, process, and technology, and then within a sentence or two, the conversation shifts to how this application or that technology will be applied. The integration of people and processes, essential to making a comprehensive solution successful, are more often relegated to the background, and the focus is shifted to a technology-centric discussion. These conversations rarely start with a holistic view of potential solutions or include the human dimensions of successful KM. One reason we believe they focus mostly on technology is the tangible nature technology offers and the lack of understanding that KM should not be about just building portals and storing data, but enabling people and knowledge flow, especially those working in teams to perform at higher levels.

Furthermore, due to the difficulty in visualizing the connection between KM and the potential for higher team performance, organizations will often rally around the information as the central object, which usually means a codified artifact, written down, stored, organized, and then ultimately moved throughout the organization using technology and information systems. And while this satisfies a portion of the job needs (some studies argue that only up to 20% of what is used on the job is codified and explicit), organizations struggle to find and apply the right knowledge, often tacit in nature, and the effective means by which they make that knowledge available and freely move it throughout the organization.

Because knowledge, by its very definition, is inherently a human quality, attempting to manage all knowledge as explicit, codified artifacts using technological systems ensures that the greatest portion of the knowledge we need to work effectively will never be available. As a result, our expectations for what we expect teams to produce will fall short. At the same time, technology cannot be ignored. Advances in information systems and improved infrastructure opened the door for unprecedented opportunities and challenges. World connectivity ensures that we can connect our workforce to deal with significant events, such as new product ideas, market fluctuations, or pandemic health issues, faster but no more effectively (and in some cases less effectively) if we lack a framework in which to collaborate and produce meaningful outcomes. The increased speed provided by connectivity cuts both ways—increasing the speed at which we solve the wrong problem or applying an out-of-date process does little to guarantee us high performance.

Institutions were constructed and designed for slower times, and their accompanying bureaucratic processes anchor us to the twentieth century. The fact that events happen faster than the rate from which they can be cognitively absorbed and analyzed or reacted to should alarm us all. The solution to this rapid evolution in technology is not just the addition of more technology, but finding the means to connect what is trapped inside of an organization's collective or networked brainpower through their knowledge and experiences. As Karie Willyerd and Jeanne C. Meister capture in their blog on *Harvard Business Review*'s Web site (Willyerd and Meister, 2009), "More companies are discovering that an über-connected workplace is not just about implementing a new set of tools—it is also about embracing a cultural shift to create an open environment where employees are encouraged to share, innovate and collaborate virtually"

(2009). The future of KM must help find a solution that unleashes the power of a connected, networked workforce over traditional, hierarchical structures, to capitalize on a new open and transparent environment. The approach must satisfy the human needs generally associated with a team that physically collocates and works together on solving problems. This team has a basic need for shared understanding of the activity and deliverables, as well as the purpose and value they will bring to the organization. They require a level of shared trust that will grow as they come to understand the competencies of other team members and what they can accomplish together, and they require commonly agreed on operating agreements to guide activities and create efficiencies. It is not uncommon to list these qualities as a goal for every team, but the issue remains how to achieve them quickly and effectively.

The emerging understanding of the power of social networks with respect to collaboration, combined with the growth in popularity of social networking sites, appears on the surface to offer a potential solution to the problem of improving teams. However, an organizational response to copy market successes by creating a corporate social networking site, or worse, jamming into an existing portal or applications, reinforced by policies to encourage sharing (sometimes seen as forced sharing) seems destined to fall short of sparking the necessary culture shift required for high-performing organizations. Clearly there is a gap, whether corporate, academic, or government organizations, in which the organizations are under pressure to maximize resources and apply innovative solutions to keep up. The key, therefore, is not a generic social network among the workforce, but to apply the power of those social connections toward an existing structure already designed to make organizational meaning—teams.

Peter Senge, in his decisive work, *The Fifth Discipline: The Art and Practice of the Learning Organization* (1990), tells us that teams are where work gets done in organizations and that successful learning organizations have disciplined approaches to forming and developing teams. Therefore, leveraging the power of more open sharing demonstrated within a social networking environment is perhaps best focused and applied within an organization in the shape of teams that form around peer relationships where everyone's value is equally recognized and valued. These teams are empowered by their social connections and armed with knowledge that has a degree of structure that is often based on little more than collaboratively agreed upon "rules of engagement" and trust in sharing with one another. These types of teams do not just happen, and they need the supportive leaders, an accepting culture, and knowledge management that result in "stimulating a social process of learning much more than merely supporting an important technical process of communication" (Bradford and Brown, 2008, p. 144).

Conventional wisdom in how to address teams with varying cultures, business practices, and different processes, which are geographically separated (often across many time zones), creates challenges to organizational sharing and effectiveness previously described. Additionally, most organizations lack the doctrine, policy, and supporting structure to aid in employing these high-performing cross-boundary teams. Based on our own observations of over 20 organizations from military, corporate, and nongovernmental organizations, many have turned to technology as the means to satisfy the requirements for sharing information faster and more effectively. However, the addition of technology alone has, in each case, produced disappointing results. Technology alone leads to information overload, a lack of clarity, and slower decision making within teams, and results in not achieving the desired team effect (Hackman, 2002). Part of the problem, according to Hackman and supported by our own observations, is that organizations fail to provide the enabling structure required for teams to work at high levels of performance.

Those organizations that address "teaming" do so in general for homogeneous and hierarchical teams; their corporate procedures provide little, if any, "how to" guidance on forming and launching diverse teams, guiding those teams through their work, and then sustaining them as membership, missions, and environments change. Applying what is traditionally understood about teams through a lens that builds on the characteristics of strong interpersonal relationships, enabled by knowledge management, and armed with technology has the potential to unleash the power of teams in a way that is meaningful and more productive within an organizational setting.

However, before continuing the discussion about teams, we need to pause and reconsider the traditional knowledge management framework of people, process, and technology before moving into how it is applied to teams. The historic framework for KM is too limiting today—each dimension is not clearly defined enough to assume that a common definition is well understood, and solutions are often generalized to a point that they fail to affect the human behaviors required to perform effective KM. Likewise, KM today focuses mostly on static knowledge rather than

FIGURE 15.1
The knowledge environment.

knowledge flow, yet we know from Neissen (2006) and Leistner (2010) that for knowledge to be useful, it must move through the organization, enabled by the KM framework. As a fundamental framework, any new approach must adapt and refine the meaning of the original three components and include additional components of organizational culture, organizational structure, content, and knowledge leadership, which we define as the knowledge environment (Figure 15.1) (Prevou, 2011). The purpose of the knowledge environment is to create a tangible approach to enable internal and external knowledge flow between and within an organization so it can learn faster and be more competitive.

THE KNOWLEDGE ENVIRONMENT

If KM is a deliberate approach to help organizations assess, create, organize, integrate, maintain, transfer, and effectively use and reuse what they know (both tacit and explicit) to achieve a sustained advantage, then what type of framework is needed? To be effective we need a framework that helps us define and manage the components of the knowledge environment, not just the knowledge as artifacts.

Knowledge is social and only moves through people. Information systems can only store and move the data and information. Information systems get information in the form of an artifact to people at lightning speed, but KM ensures it is the right information going to the right people at the right time, to enable them to decide and act. Effective KM requires high human-to-human interaction and helps eliminate barriers to knowledge flow by networking the hierarchy of an organization. KM is a discipline that treats intellectual capital, both tacit and explicit, as a managed asset, whereas information management systems manage just the explicit documentation. The KM discipline is more holistic—knowledge managers should strive to manage the *knowledge environment,* not simply the assets, to produce some type of measurable outcome for the organization. The *knowledge environment* consists of seven major components: organizational structure, people, processes, technology, content, organizational culture, and knowledge leadership.

The people, processes, and technology intersect, forming linked components that must be in balance. Culture, content, and structure are independent variables that affect each of the linked variables. Knowledge leadership is overlaid across all the components and provides the vision, drive, and resources to make KM possible and effective.

> For knowledge to be effective, it must flow. Successful flow demands a system that can *push, pull, and prod* users.

Furthermore, knowledge must be aligned with the organization's business process so that the movement of this knowledge is *pushed* by experts who know the type and value of information necessary for a process, activity, event, or project. Knowledge can also be *pulled* by team members when they know what to look for to satisfy an immediate need. Finally, knowledge can flow by being *prodded* by other experts and intelligent systems to those who need it. For example, Amazon.com* sends you an e-mail and prods you with knowledge about other people who have bought book X and were also interested in book Y, therefore making knowledge available for use based on predicted need. In the end, the most effective social network systems include all three push, pull, and prod approaches. Effective KM environments enable the push, pull, and prodding of mission critical knowledge.

This integrated knowledge environment is an ecosystem that requires a balance of three types of interactions: human-to-human, human-to-system, and system-to-system. These interactions are critical to an organization's ability to function properly, and the organization's knowledge leadership, structure, people, processes, technologies, and culture make it possible for the "flow" of data to become information and then knowledge. The paradox is that knowledge does not flow naturally within our complex organizations. Barriers come in all shapes and sizes, and KM cannot be left to happenstance if we are to stay competitive and continuously learn, innovate, and adapt. Knowledge leadership is required at the highest levels of an organization to resource, prioritize, and advocate for a deliberate KM approach. Knowledge leadership at the middle and bottom of the organization is also required to innovate, identify opportunities and threats to knowledge flow, and practice effective KM strategies to prevent the loss of organizational knowledge.

The knowledge environment must address the full spectrum of knowledge—from explicit knowledge that we can write down and manage as a physical artifact, to the more elusive tacit knowledge (the knowledge in our heads), which many argue cannot be made explicit. However, through the proper techniques, it can be brought forward, made visible, and shared in some limited fashion. Without understanding the difference, we try to manage all knowledge as explicit, falsely believing it can be captured, stored, and shared through electronic means. Nothing could be further from the truth, and this accounts for the frustrations most organizations experience, given a heavy technology emphasis.

The traditional knowledge management framework (people, process, technology) emerged from the information management movement in the 1990s at the dawn of the Internet and the reduction in costs to collect and store data on a massive scale. Today, advances in technology and global connectivity push KM into newer, uncharted waters. The successful transition from an information management to a knowledge management environment is complete and must now give way to a newer, more expansive view of the framework into one best defined as an operating knowledge environment. This adds four new components that account for culture, structure, content, and leadership as they relate to KM.

By orienting on managing the knowledge environment rather than the knowledge, we make the components of the knowledge environment tangible. Each has form, specific inputs, outcomes, and subprocesses, and can be measured more effectively. This framework applies a systems thinking approach to KM and requires each component of the environment to be in balance with the others. The exact quantity of each to maintain the balance will change as environmental conditions change. The art to managing the knowledge environment is the balance, while the science is oriented on the tangible human behaviors desired as an outcome. The traditional KM model of people, processes, and technology being insufficient, we briefly attempt to describe each of the seven components of the knowledge environment as we see them.

People

> One way to adjust our thinking about people, process, and technology is to compare KM to a project, like building a house, and the effort to grow a garden full of fruits and vegetables.
>
> With a project, such as a house, we can identify objectives and timelines and predict the number of workers and needed supplies. If we fall behind, we can add more workers or supplies to build the house faster.
>
> In a garden, plants grow as fast as they grow. Gardeners can water the plants and fertilize the soil to encourage growth. If the plants do not grow as fast as we like, adding more fertilizer, water, and gardeners will not make the plants grow but is very effective in making a sloppy mess.
>
> Because people are ultimately the center of any knowledge management initiative, we will forever be stuck growing a garden. Throwing too much process and technology at a problem will not make people share faster than their culture will allow, but it is effective at making a more expensive mess.

The traditional knowledge management framework for people includes everything from the people and their attitudes toward sharing and collaborating within the boundaries of the cultural context in which they live and work to their training. These characteristics and human behaviors are essential to organizational success, but when lumped together as a single "people" component, they become diffused and diluted. When we try to develop a KM solution, it then becomes impossible to distinguish which characteristic or behavior to orient to affect a positive change. When we speak of people, we refer to their ability to understand, learn, build relationships, communicate, and collaborate effectively, and apply the processes, technologies, principles, and strategies needed to perform their work effectively. The knowledge, skills, and abilities of each knowledge worker must be deliberately developed. We do not send soldiers into combat without proper marksmanship training, yet we often send them into staff work and combat without training on how to form and operate teams or use the knowledge management tools and processes.

Culture and attitudes play a key role in how fast and effectively knowledge will ultimately flow and are more important than employing powerful technology and efficient processes. In the end, people connect with people based on their relationship with one another. People do not connect directly with technology, although this may seem the case when observing someone blindly focused on a mobile device, but rather connect virtually to others through technological means. This sounds intuitive and appears obvious, but many times analysis stops at the human–machine interface without thinking through how connections between people are made via a system, whether synchronously or asynchronously. For example, even something as explicit as this chapter within this book is ultimately an asynchronous discussion between the authors and you—the reader—with the book only serving as the enabling tool. Applying the human-to-human interaction through technology is critical to the success of any knowledge management framework, but essential to applying it at a speed and depth to make such tacit sharing meaningful. The person-to-system interaction (often the usability of the interface) is only important after you understand what people-to-people interactions are needed.

Process

KM PROCESSES

- Knowledge planning
- Knowledge creation
- Knowledge acquisition
- Knowledge integration
- Knowledge organization
- Knowledge transfer
- Maintaining knowledge
- Assessing knowledge

Process in the current model is not well defined except that the model generally covers any or all of a myriad of different types of process.

This category is most often defined in one of three ways: as a systems approach as highlighted by the Army in their KM Field Manual (FM) 6.01.1 (U.S. Army, 2008) where "The KM Process" (as if there is only one) describes a model with four steps—assess, design, develop, and implement. This basic systems approach can be applied to most anything we build or create as a framework of KM processes has proven virtually useless without additional context. The second "process" often described is the business processes, or the activities that make an organization function. Last are the KM processes needed to acquire, create, organize, share, and use knowledge effectively. We believe this last set most clearly represents the "processes" component of the knowledge environment.

When referring to the KM processes in the knowledge environment framework, we are talking of the processes that help us convert knowledge to action and achieve results. These are not the organizational business processes. These processes (and there are many) have outcomes that contribute to learning and accomplishing organizational objectives. A variety of knowledge processes exist based on who you read. All have some degree of knowledge acquisition, organization, and dissemination. Most KM literature cites knowledge *use* as central to each KM process rather than as a process in and of itself. These processes apply to individuals, organizations, and across the enterprise. Failure to understand the KM processes and address them directly is one of the major reasons stovepipes continue to appear and we have difficulty sharing across organizational boundaries. Technology will enable these processes, but in the end, the ability to share and use the knowledge is a human endeavor.

Technology

Technology is perhaps the best understood of the three characteristics in the traditional model and usually focuses on specific software and applications that will enhance storing, accessing, and retrieving information, as well as synchronous and asynchronous collaboration.

Because of the lack of defining the people and process components and the interactions that make knowledge flow, technology has been looked at to solve knowledge-based problems. Fixing technology and application issues often becomes the central focus of knowledge management initiatives because there is more clarity in identifying tangible goals and metrics, yet they often do not produce the desired positive return on investment. If we want to change human behaviors and make that change lasting, we need a more refined model that facilitates understanding the gaps and problems so we can create practical and effective solutions. Technology enables KM; it is not KM. Technology allows us to reach further and span time and geography. It also allows us to store and move unthinkable amounts of information and data. Too often we have seen organizations equate Microsoft® SharePoint® to KM. Technology is essential for effective KM, as we can no longer rely on sharing with just a local group as we did in the past. We can manage the technologies and access them either locally or across the enterprise.

Every organization we visited has a deficiency in information technology (IT)/KM technology training, primarily because some leaders continue to believe systems are intuitive or that training provided once is sufficient to sustain you for the next few years. However, the most successful organizations have continuous training on systems and processes available for all workers, not just for "knowledge workers." Technology in support of the knowledge environment must provide a suite of services and applications, for both synchronous and asynchronous interactions, and must provide the security protocols required to protect information without severely restricting access to critical knowledge. On the other hand, much of KM requires no technology.

Organizational Structure

Organizational structure refers to each specific organization and will differ from one organization to another. This component includes the organizational layout, the governance and controls applied, the business processes, the infrastructure, and the physical arrangement of where people sit and who they interact with regularly. In one organization, knowledge flow improved by over 30% simply by changing the seating arrangements of people in the office.

Much of the component "organizational structure" is about human-to-human interactions and removing the barriers that inhibit these interactions. In the traditional KM model, the application of organizational structure was diluted in the "process" component and often orientated toward the information that will ultimately flow through the workflow rather than analyzing how people will connect with each other and into a specific process. The information-centric view makes sense when applying technology first, but if we reverse the framework and think of how a team forms around a process and evaluate the knowledge flow from a people-to-people and then a people-to-system interaction perspective, then process efficiency becomes easier to define and makes applying technology more effective. When this approach is used in complex collaborative environments where processes are applied to people before technology, simple solutions are often identified that are already on hand rather than acquiring expensive technology. With respect to governance in this component, the strategies, rules, and procedures an organization puts in place one day may not support the rapid collection and sharing requirements of today. As part of our analysis of this component, we always ask if rules are first still useful or needed, then if they are addressing the right processes and interactions, and enabling technologies before we attempt to adjust or redefine the rules.

Content

This component consists of both tacit and explicit knowledge. According to many recent studies, 80% of the knowledge we use on the job is tacit, the majority of which cannot be made explicit, yet we continue to expend nearly all our resources on technology solutions that can only organize, store, and move 20% of what we know or need on the job. Many KM initiatives failed to even identify the tactic knowledge required and generally disregarded the ability to manage tacit knowledge. Most KM programs fail to understand that knowledge flow requires movement of both types of knowledge to show value. Our IT systems are great at storing and moving data, but information at rest in these systems is virtually useless. The more we share and adapt it, the more valuable it becomes. Making tacit knowledge flow requires person-to-person interactions and collaborative sessions to assist content flow.

When content is considered as part of the people or process component in the traditional KM model, it is almost always thought of as artifacts, documents, and media rather than expertise and relationships. We believe that content is too valuable to leave to happenstance and that it must be addressed as its own component within the knowledge environment.

Organizational Culture

This component is about managing behaviors and values. It is about creating an obligation for continuous learning and sharing. "Knowledge is power" is an outdated axiom and can often cost lives or revenue. The new mantra for a high-performing learning organization must be about the power of knowledge shared to create a culture of collaboration. To effect change, we have to focus on specific behaviors, and without understanding this component of the knowledge environment, most change initiatives never live past the leader's departure.

Culture change can and must be deliberately and delicately managed. It is too often the crutch upon which KM practitioners wring their hands about KM project failures. Organizations must set the conditions for the people to change the culture in order to share in a way that is better aligned with how we think, solve problems, and interact with one another. To address culture, we need to focus on the behaviors we want workers to perform and then create the incentives and forcing functions to shape that behavior. In many cases, a failure of some to adapt to a new process or technology insertion is seen as a culture problem, when in fact it is usually related more to the individual's training, aptitude, or willingness to adapt.

Knowledge Leadership

If all other components of the knowledge environment are perfectly balanced but no effective knowledge leadership exists, the frustration and lack of resources will quickly grind any KM initiative to a halt. Knowledge leadership does not have to come from the top. In many cases, we have seen it from the middle or a grassroots level. These efforts are often slower and fail more frequently as they are crushed by the bureaucracy. More often than not, the obstacles to effective collaboration also come from the middle levels of our organizations. They are the ones who do not understand a systems approach, the speed or complexity of knowledge flow, and are too busy to learn or get organized. Good knowledge leaders do not need to use every system. In fact, as we visited leaders from over 20 organizations, few personally use the tools; they have staff that use them and that's where the real value comes in. Leaders should demand the tools be used; rather than briefing from Microsoft PowerPoint® slides, why not brief directly from the organizational knowledge system? Instead of planners developing a strategy or proposal and e-mailing it for changes and concurrence to 50 different people, why not demand using collaborative tools and build it online together? This is what is meant by setting the example, not being an expert in Microsoft SharePoint.

As a result of talking with leaders, here are few things they can do to improve collaboration and KM implementation in their organizations.

Effective knowledge leaders will

- *Make collaboration and sharing a top priority and put it on the agenda.* Have you ever been in those meetings where the staff only briefs the boss? This means dedicating part of every meeting to the sharing of ideas and innovation. Often, the leader must facilitate this exchange. Good knowledge leaders dedicate a third of their meetings to cross-boundary sharing.
- *Establish and communicate a knowledge vision*, allowing the organization to
  - *Manage conversation.* Make sure we are collaborating up from, not on the back end of, the project or a process.
  - *Enable knowledge activists.* Seek out the people who get it and let them run the ball.
  - *Globalize knowledge.* Make sure critical knowledge is visible or available to everyone in the organization. This is as much about transparency as it is about knowledge transfer. If Amazon.com can do it, why can't every knowledge system?
  - *Manage change processes.* We have deliberate processes for most everything we do, why not a process to manage change? It is too important to leave to happenstance.
  - *Develop knowledge leaders in the organization.* This is the middle level of the organization and the ones who set the tone for the culture of collaboration. As they grow they must be thinking "who else needs to know this" all the time. Again, shine the spotlight on them when they effectively enable knowledge flow. Development requires education and training to make them successful. The number one reason most people do not adapt is they do not have the knowledge and skills required to operate within the new process or use the new tool. Create an expectation of continuous learning, and promote those who do.
- *Build a guiding team for KM initiatives.* Find the early adopters and put them on the team. Shine a spotlight on their efforts and reward their successes. Celebrate their failures and identify the lessons learned from those failures and share them.
- *Create an obligation to share.* Make people accountable for sharing. Squash those who hoard knowledge; reward those who share and learn from others regularly. To establish a culture of collaboration means changing behaviors.

- *Enable action by putting tools in place.* Tools do not equal technology here. Resource the right people, the right processes, and the right tools to enable sharing. Technology is not always the answer; sometimes it is more effective meetings and working groups or where we bring people together to collaborate during a process.

Leading and managing the components of the knowledge environment should be no more difficult than leading and managing other functional areas. Everyone should be practicing KM, but the leaders set the tone and the priorities. If what the boss checks is important, then maybe doing a few of the items listed here will create momentum and lasting change.

Managing the knowledge environment should not be left to chance. We do not build schools and expect learning to occur without having teachers. We do not build libraries and expect the patrons to understand the filing system, manage the stacks, or know where each type of book might be located. For these reasons, dedicated KM professionals are needed in every organization to assist the leadership in developing plans and policies that govern human capital, integrating and training personnel, managing tools and content that facilitate the human-to-human and human-to-system interactions, and programming the system-to-system interactions.

The knowledge environment is a framework that we can manage. It offers a practical approach to a discipline saddled by heavy baggage and misunderstanding. We must create a culture of collaboration and knowledge sharing in organizations and teams in which not only is key information "pushed and pulled," but organizational prodding helps connect the workforce, build relationships, and access the global knowledge needed now to meet mission objectives.

APPLYING THE KNOWLEDGE ENVIRONMENT TO TEAMS

Knowledge management is most often seen as just a technological solution, and most organizations do not address ways to build the KM-related knowledge, skills, and abilities required for effective, high-performing teams. Rather, they focus on technical expertise and assume teaming skills exist. With respect to how we see and apply knowledge management to teams, we must have a broader approach and look at ways to manage *all* components of the knowledge environment (Prevou, 2011) and keep all seven elements of this environment in balance as with any ecosystem. We believe one of the greatest frustrations in KM has been the lack of understanding of how the components of this ecosystem must work in harmony with one another to produce a desired result. The social networks formed by any organization are integral to how the organization will move knowledge through the organization, and each component of the knowledge environment will add value to the flow of knowledge. These social networks (we are not talking Facebook and LinkedIn here) are the connections and relationships across boundaries of even the most structured hierarchical organizations that in effect "work around" bureaucracy and hierarchy determined to create efficiencies. The knowledge environment for a team must support these natural human behaviors and relationship requirements.

With respect to the teams, rarely does an organization's doctrine or policies address building an organizational culture of collaboration or creating an integrated virtual work environment that would underpin the effective forming and employment of cross-boundary teams. Instead, teaming is implied or left to luck (Lipnack and Stamps, 2000). A structured approach, therefore, that combines governance, method, process, and technology is critical in not only developing these high-performing teams, but also in lubricating the relationships and processes that enable knowledge flow to achieve higher performance faster to ensure work gets done in time to make a competitive difference.

Knowledge management enables both formal and informal team structures, yet only a few organizations we have seen have formally acknowledged and attempted to harness the power of KM and informal learning to enable teams. We believe these few organizations truly understand what Senge was trying to describe in his learning organization. If knowledge is inherently a human quality and people are social, then knowledge can only move when people connect with one another. Connecting people together and calling them a team is far short of the discipline required for an effective team. Many studied teams and their effectiveness, and we believe understanding what makes a team work—and not work—is critical to achieving high performance. We also believe that enabling the teams in a way that makes the most sense and ensuring their team knowledge environment exists and is in balance is key to high performance. Before we discuss the approach we piloted to create more effective teams faster, let's look at teams and what makes teams work.

ENABLING TEAMS

If teams are the cornerstone for a potential organizational strategy to take advantage of the connections across a social networked and connected workforce, then why has a cultural shift not already occurred? Furthermore, what are the characteristics of a good team, and can a sound KM framework in fact enable them to be more effective and higher performing than traditional teams? Generically speaking, the belief that working in "teams" makes us more creative and productive is widely held by organizational leaders who are quick to assume such teams are the best way to get results; however, many times team performance falls short of expectations.

Research shows that although the formation of teams allows for the introduction of more diversified talent, they will still underperform despite the addition of additional resources (Coutu, 2009). Forming teams alone does not guarantee success or higher performance. Richard Hackman, a recognized expert on teams, in his book *Leading Teams*, outlines five conditions that must exist for teams to be successful (Hackman, 2002, p. 31):

- An understanding of who is on the team
- Compelling direction or purpose
- Enabling structure
- Organization support
- Expert team coaching

Hackman suggests the failure to ensure these conditions are present requires us to rethink the importance of teams in organizations and how they are enabled and resourced for success. However, teams can achieve high performance if they have a structured team formation process and an approach that is enabled by knowledge flow with access to critical information. Key here is that organizations need not rethink the *role* of the team, but the process of teaming and how to develop the team and then arm it appropriately.

Providing a supportive knowledge environment to enable high performance in teams is often an afterthought in many organizations. Applying the five characteristics defined by Hackman within the unique circumstances of the organization is almost nonexistent in any of the knowledge management programs we observed. For example, "We spend millions on individual and collective training" was noted by a staff officer working within a large military headquarters, but,

> We assume we develop teams well, and the truth is we do develop our hierarchical unit teams well. But those teams are from our same culture, training background and wear the same uniform and share a common language. Where we don't do as well is when we are forced to team across organizational, service, interagency, or multinational boundaries. (Prevou et al., 2009, p. 2)

Because diversified teams will require participation from across various cultures, organizations, or even with different levels of command or structure, they will all view knowledge and collaboration differently. Team members will often talk past one another despite agreeing on the problem, strategy, or way ahead simply because they approach a problem from various points of view. Such conditions are further compounded because diversified teams bring different experiences, expertise, or backgrounds to the group. Technology allows for the expansion of teams, adding a geographically dispersed connection with many of the team members. This allows for broader skill sets to be included in tackling problems, but at the same time, risks compounding the problem in establishing cohesion among participants and growing the team too large to be managed effectively. As a result, often the relationships and the ties that help us understand who is on the team and why they are there are not well developed. Likewise, the ability to develop a shared vision of the future is often hindered by the time and space that technology easily spans. Couple that with the slower development of relationships and we start to see the team life cycle slow to a crawl; knowledge does not flow and productivity stagnates.

Few organizations understand how the team must use all the components of the knowledge environment at the outset of the team's life cycle as highlighted earlier. There is general misunderstanding on what tools, systems, and processes are required to create the environment that all team members will have the appropriate level of access and skills to navigate. They appear equally ignorant of (or blind to) the need to create positive conditions for knowledge flow, and rarely is there recognition of the need to make the investment in a coach who can stand back and focus on the team process and facilitate, rather than lead the team. As a result they are satisfied with the team lumbering along or they hold team members responsible for the infrastructure gaps for which they have little to no control over. Teams may eventually overcome this challenge, but as Hackman implies, such achievement is an assumption, not an inevitable fact. In a high-speed competitive environment where millions of dollars are spent on technology and processes toward knowledge management programs, trusting luck does not generate a confident return on investment and higher performance is not achieved fast enough to make a difference.

Hackman's teams seemed to have certain homogeneous qualities, but the types of teams we observed come from many organizations, countries, and political and functional backgrounds and most often do not have a designated hierarchy. They appear to act more like social networks than organized units that are cross-boundary and often geographically dispersed. These teams have primarily been U.S. government interagency and military teams and public healthcare teams consisting of all walks of medical practitioners, community health professionals, local politicians and government employees, citizen advocates, and healthcare volunteers with no authority over one another often competing for the same resources. For this reason we call them *leader teams* to distinguish them from more traditional hierarchical teams that are better understood. These leader teams do not live and work with each other daily, so developing relationships, working agreements, common purpose, and trust must be deliberately developed.

These teams consist of leaders who come from different organizations, often with no formal authority over one another, who often come with differing criteria for success, often with different operating procedures and work ethics, and almost always with no common knowledge system where they can share and collaborate quickly, effectively, and freely. Each brings to the team a unique set of skills and each has the reach back into their respective organization and the power and authority to make decisions and commit resources or directly call someone who can.

The need to establish solid knowledge management support for these teams is an essential part of the enabling structure and organization support mentioned above. Rather than making the creation of a team knowledge center part of the team startup process, companies assume that critical knowledge related to the team's objectives or goals is available to the team and will naturally be understood the same way by all team members. The assumptions take the place of conversations about how the team should operate. Knowledge management in the case of the team must consider many factors, including the cultural aspects of each team member. This consists of more than national or ethnic cultures—but the biases and values participants bring to the team. Attorneys think differently than engineers who look at the world (and a problem) differently than accountants and the marketing department. Knowledge management must transform from an artifact-centric view that is organized around documents and technology to one that is centered on people organized in a team that is enabled by knowledge.

Social networks are one approach to ensure knowledge flows through the interconnected teams. Appling this approach against a team construct allows organizations to leverage the power of relationships in a meaningful and focused way. The ability to stay connected to an ever-larger network is outpacing traditional work processes and social norms that bind team members together. Today, with the explosion of readily available mobile devices, this trend is more likely to grow. Generations X and Y are well ahead of the baby boomers in their acceptance and application of collaborative technologies, but they often lack the relationship-building skills necessary to work in a diverse team that comes with time and experience. One reason these skills are underdeveloped is because of the haphazard way organizations go about forming and launching teams. The result, once they have formed, is that most team members go off and work their issues/problems/task independently rather than in collaborative settings. They return with half completed documents or report progress from a single point of view versus using the collective wisdom of the crowd earlier in the process to stimulate thinking and innovation.

Because the diversified teams do not normally have preestablished relationships and commonly used knowledge systems, their greatest strength—their diversity—contributes to a lack of shared purpose and the increased confusion around why am I here and how will we operate together. When knowledge is not enabling, but disruptive, causing a breakdown in communication and effective collaboration (often described as the storming experience, where team members are defensive and protective of their ideas and parent organization), the challenge to move beyond is prolonged and often frustrating, which inhibits high performance. Adding mobile technology and expecting the team to be connected continuously will not lessen the burden or overcome the need for interpersonal relationships, nor will it alone reduce the length of the storming phase in team development. On the other hand, introducing mobile technology to a team with strong bonds means access to knowledge is no longer confined by space or time, making rapid and responsive decision making possible.

Because, according to Hackman, teams underperform when the basic conditions are not in place, the lack of doctrine, formal procedures, and supporting structure for developing teams further inhibits quickly achieving higher performance. This phenomenon was observed in the military, the U.S. government, and healthcare industry where highly specialized expertise and spiraling costs make teaming essential. While organizations like the military address "teaming" in general for homogeneous and hierarchical teams, they provide little, if any, "how to" approaches to form and launch a diverse team of leaders, guide them through their work, or sustain the team as membership, missions, and environments fluctuate. Current military doctrine, let alone that of most industries like healthcare, does not address ways for teams to build the skills, knowledge, and abilities required for effective high performance. Nowhere does military doctrine address developing high-performing leader teams.

Of the 120 top teams Hackman and his team researched, almost all agreed they had set unambiguous boundaries. Yet, when researchers asked individual participants from among the teams to describe their fellow team members and their capabilities, less than 10% agreed about who was on the team and why. Hackman goes on to say that teams must have a clear, compelling direction and a shared purpose, but notes the difficulty in creating that shared understanding of direction or purpose. As Hackman discovered, larger teams increase the number of links to be managed, and maintenance of those links causes team performance to break down (Hackman, 2002, p. 45). We witnessed these symptoms in each of the more than 20 teams observed over the past 2 years. Some of those observations are as follows:

- Teams do not know the other members of the team and what they bring by way of capabilities.
- The members of the team rarely agree on the team's purpose.
- The members of the team often understand what must be delivered, but differ in their idea of approach and content.
- Some team members drive toward producing products rather than developing relationships.
- The teams lack commonly agreed upon operating agreements.
- The teams lack a common knowledge system to communicate and collaborate virtually.

- The teams lack an understanding of the network each of the other members has available and how that network can be leveraged to solve this team's problems.
- Most teams do not collaborate early in the team's development or in solving a problem, and time is often wasted on tangents and misdirection.

DEFINING HIGH-PERFORMANCE TEAMS

One of the challenges we struggled with throughout our work was defining high performance for the teams. Standards vary widely, and in some places, high performance was getting everyone to communicate in a civil manner for the duration of a meeting. In others, it was about producing a product in a very short amount of time that would determine the continued existence of the organization. Each time we start a seminar or workshop, we ask the team to define success and then ask them to define high performance. Very often they are surprised to see that the two are not the same and that with a little structure, collaboration, and coaching, it is easy to achieve their version of high performance. There has been a great deal of work done by others on teams and defining high performance. We hope to build upon their work and provide you some metrics that will be valuable as you develop your own teams or participate on teams. Our list of leader-team characteristics developed over time rather than as a result of any survey or formal study, and undoubtedly some characteristics are included in the work of others.

As intergovernmental agency missions expand to include lead and supportive roles in complex political and conflict environments, so does the recognition of a vacuum in the U.S. government's ability to effectively form and launch the diverse teams we are discussing. General Martin Dempsey's remarks to the 2009 Joint Warfighting Conference (JWC) define the challenge of the next 20 years in terms of Joint, Interagency, Intergovernmental, and Multinational (JIIM) networks. In circumstances such as Hurricane Katrina, there may be no "lead" agency at all, requiring each team or agency to adapt to the other while organizational leaders must rapidly come together in tandem to form cooperative teams. As General Dempsey states, "If we are to be truly committed to becoming a Joint Interagency Intergovernmental and Multinational Team, then our interagency and coalition teammates are going to have to match our decentralization of capability and decision-making authority with their own" (Dempsey, 2009). Implicit in General Dempsey's address is that constructive relationships among agencies, industries, and nations are central to the success or failure of the decentralized joint, interagency, intergovernmental, and multinational network.

The evolving U.S. Army Mission Command doctrine acknowledges the need to synchronize information technology (the network) with knowledge management to create a culture of collaboration. It describes the need to develop "the cognitive ability in leaders to master transitions, innovate and adapt" (Leader Development Strategy, 2009, preface, p. 1). Although the new doctrine captures the challenges required to solve complex, ill-structured problems and work across boundaries, cultures, and organizational hierarchies, it suggests a methodology is still needed to develop the knowledge, skills, and abilities required for this type of teaming.

High-performing teams make an impact. They can be counted upon to get things done. They adapt quickly and have an increased sense of shared vision and purpose. They have a strong sense of trust for one another and understand the competencies that each member brings. These qualities seem to increase team confidence. High-performing teams not only communicate more effectively, they collaborate and build things together more often and more effectively. These teams build relations and leverage the talent of not only the team members, but the people known by the team members—their Rolodex. High-performing teams see obstacles and work around them. They are not hindered by hierarchy or bureaucracies, and they form strong social networks. They develop operating procedures to guide simple and complicated processes so that their energy can be focused more effectively on complex problem solving. High-performing teams share knowledge as well as information and use technology to expand their reach and increase the speed of collection and dissemination. They feed each other's creativity by challenging and questioning in a respectful and professional way, and they support one another once the relationships have been built. High-performing teams are also quick to cut out the unproductive "dead wood" and isolate them from the group.

Of the teams we observed, those that were rated as "high-performing" by peer teams and superiors had higher levels of these qualities.

Characteristics of a high-performing leader team are as follows:

- Make an impact
- Are adaptable and flexible

- Share cognition
- Share vision
- Share attitudes/trust/confidence
- Communicate meaningfully more often
- Share both what they know and what they do not know
- Collaborate frequently
- Build constructive relationships and bring others into the team
- Are undeterred by bureaucracy and work around obstacles
- Develop operating agreements to govern how they work
- Use technology
- Feed each other's creativity
- Challenge and support one another
- Do not tolerate dead wood

This is a long list, as most organizations identify only five to six characteristics at any one time. Our belief is that true high-performing teams display all of these characteristics more often than average teams.

A NEW APPROACH TO BUILDING LEADER TEAMS

A new approach we call *teams of leaders* (ToL), was the evolution of previous work surrounding commander leader teams initially conceived and promoted by Lieutenant General Frederic Brown (Bradford and Brown, 2008, p. 78) within military units to expand the breadth of collaboration and sharing beyond traditional organization boundaries. The approach was applied by the European Command (EUCOM) starting in 2007,

> To build strong relationships with long-standing allies and to build new ones with emerging partners. For example, in addition to traditional missions like humanitarian assistance, other challenges also exist—such as the flow of energy, financial turbulence and threat of pandemic disease—that require innovate expertise and experiences to mitigate them before a crisis develops. (Brown, 2009, p. 1)

ToL was an initiative by General John Craddock, EUCOM Commander, against a number of staff directorates within the headquarters over the course of several years. In winter and spring 2009, EUCOM expanded the approach to the nested interagency-military teams required for whole-of-government interoperability during an operational crisis that was eventually captured in a Teams of Leaders Coaching Guide (2009). General Craddock, at the end of his tenure in 2009, described ToL as follows (Brown, 2009):

> During my tenure as EUCOM Commander one of the two most significant "wins" was the Command's embrace of the Teams of Leaders concept. Without question—ToL was and remains the enabler for a significantly higher performing staff, increased horizontal and vertical communications, and shared priorities and focus of effort. This—ToL—is no silver bullet—not fairy dust—but rather the application of enlightened, thoughtful, effective procedures by talented professionals—commencing with a series of "ah-ha's" that quickly become self-generating. (General Bantz J. Craddock [U.S. Army, Ret.] Commander EUCOM/SACEUR 2007–2009, p. 7)

During a major exercise and in response to a crisis in the Republic of Georgia during the summer of 2008 (Hilton, 2009), ToL demonstrated the ability to significantly improve the team's understanding of a situation and its shared purpose and vision for an operation within a diversified, but interconnected knowledge environment across Europe. Through this relationship-building process, ToL demonstrated the ability to improve trust and help develop the sense of shared competence and confidence required for complex mission sets among diversified groups and rapid knowledge sharing, while incorporating greater numbers of people and organizations.

Defining Teams of Leaders

As outlined in the Teams of Leaders Handbook (U.S. Army, 2009, June), cross-boundary teams today consist of leaders from different organizations brought together to leverage the expertise, experience, and resources of their entire organization. The most common reference in the military is the term Joint, Interagency, Intergovernmental, Multinational (JIIM) operations, which bring the full capacity of the U.S. government and other actors on the modern battlefield together to resolve an issue or challenge. In this situation, where a single authority and a single shared purpose may not be present, leaders cooperate with one another based on collective agreement to support a higher goal and the collective cooperation in developing the solution. These teams of leaders must build common understanding and relationships, and develop trust within these types of teams while building common operating procedures that help the team operate more effectively. Leader teams use many approaches to enable continuous collaborative operations as well as team building, and enable interagency reach (nested teams) as they make and execute decisions and globalize what they learn.

What has become clear is that most organizational work is accomplished in teams. Less clear is that many efforts are now being conducted in *leader teams.* Groups of action-oriented decision makers from multiple disciplines, functions, or organizations who come together to accomplish a specific purpose are nesting themselves into larger efforts to maximize capability and problem solving. Generally, leader-team members are not bound by an explicit hierarchical structure but retain the authority to reach back to their organization of origin and generate action. The leader may be an organizational representative or the leader of an organizational team, but each brings a unique contribution to the effort and acts as a decisive driver for the functions they represent. They are linked together through purpose but often lack clear lines of authority and an effective means to collaborate across time and space.

Additionally, in today's fast-changing environment, teams are frequently nested into a larger network of effort and expected to share knowledge and expertise. However, their inoperability is often spotlighted as the natural by-product of stovepiped biases, cultures and regulatory systems, geographical dispersion, methodologies, inconsistent languages or lexicons, and juxtaposed organizational interests. These boundaries hang as an albatross to interoperability in sharing essential knowledge and achieving high performance. Consequentially, friction and stagnation reduce agility and ability with, at times, deplorable results.

Nowhere was this more starkly realized than the local, state, and federal response to Hurricane Katrina. As the Bush Administration's 2006 critique of the federal response notes,

> At the most fundamental level, the current system fails to define federal responsibility for national preparedness in catastrophic events. Nor does it establish clear, comprehensive goals along with an integrated means to measure their progress and achievement. Instead, the United States currently has guidelines and individual plans, across multiple agencies and levels of government that do not yet constitute an integrated national system that ensures unity of effort. (The White House Archives, 2006, p. 66)

This however, was not caused by a lack of passion or desire to perform among participating agencies or in the recognition that more training is needed:

> LESSON LEARNED: The Department of Homeland Security should develop a comprehensive program for the professional development and education of the Nation's homeland security personnel, including Federal, State and local employees as well as emergency management persons within the private sector, non-governmental organizations, as well as faith-based and community groups. This program should foster a "joint" Federal Interagency, State, local, and civilian team. (The White House Archives, 2006, p. 73)

The lessons learned articulate the need to establish leader teams in response to adversity, and policies, laws, and the addition of technology would solve the teaming and collaboration challenges that are uniquely human. Without the shared team qualities necessary for an effective team, these additions will simply increase the speed of failure.

The challenges of Katrina were not anomalies. They were, and in many cases remain, inevitabilities. Rather than address the outdated processes, too often the solution has been to create additional parent bureaucracies that pull decision making to distances devoid of situational context. Decisive action is further constricted. These impediments dominate the effort with such efficacy that the notion of governmental unity and efficiency becomes the exception while functionality is underwritten as exceptionalism. However, the knowledge necessary to solve the problems was most likely present but trapped in silos of well-intentioned, hard-working responders willing to share, but unsure with whom to share their knowledge. Understandably, the ability to anticipate, plan, and resource for such a cataclysmic event is arguable. The range of potential crises to the city of New Orleans alone, much less every city in America, is nearly infinite.

The Three Components of the Teams of Leaders Approach

- High-performing team qualities
- Knowledge management
- Information technology

That is not to say that we are helpless—the knowledge to react will always be collectively available, and the strong desire to share, even if only at the local level if the underlying means to unleash it is provided. Subscribing to a few basic conditions for productive teaming can improve performance and solve complex, ill-structured problems with relative ease. A model that allows organizations to learn, adapt, and innovate faster is needed to succeed in the high-risk whirlwind of twenty-first century dynamics. Contained within social networks is the knowledge necessary to overcome these challenges. Managing that knowledge in anticipation of an event is impossible, but creating the conditions where participants can rapidly transform networks into informal or formal teams to respond to such challenges is achievable. The sharing of critical knowledge will then flow.

To be a geographically dispersed high-performing team of leaders requires four elements—the presence of what we call the high-performing team qualities. These include shared vision and purpose, shared trust, shared team competencies, and a shared confidence. Sound knowledge management practices and technology allow the team to connect, communicate, and collaborate effectively in both synchronous and asynchronous ways.

When the conditions have been set correctly for the development of teams, remarkable advantages can be recognized. In our most recent work, surveys of four distinct types of cross-boundary teams consisting of military and civilian organizations recognize that teams skilled in such environments display marked abilities over those that were not: Team members' confidence rose by 52%; the ability to collaborate effectively increased by 75%; clarity in mission improved by 45%; productivity increased by 25%; turnover (those voluntarily departing the team) decreased by 25%; and time to team competency increased tenfold (Prevou et al., 2011). People like to be part of "the team," but only when that team functions with a high degree of shared purpose, trust, and competence. Success builds confidence and empowers the team to confront and solve more difficult challenges. In simple terms, applying sound teaming principles and practices is a cost-effective way to increase collaboration, profits, and productivity by tapping the contributions of a broader pool of talent, their expertise, and knowledge.

The Teams of Leaders Approach

The Teams of Leaders (ToL) approach is a leadership framework for rapidly building and effectively employing cross-boundary teams that are highly competent both in making and executing decisions and in learning and adapting together. These teams are further enabled by knowledge and information management.

The ToL approach has been piloted in the U.S. Armed Forces and in programs within the healthcare field. This approach has been proven as an effective way of overcoming common team dysfunctions faster and more effectively than allowing that formation to occur naturally. It provides a specific methodology to help teams *form and launch* to achieve results more quickly, then sustain that momentum using a problem identification and solving framework. It is especially useful in complex, adaptive situations, where

- Team members come from a variety of organizations.
- Team members are spread out geographically.
- Teams are solving complicated problems, involving many issues and stakeholders.
- Teams must adapt to rapidly changing conditions.
- Teams need to multiply their effectiveness through new team-based approaches and high-tech, collaborative tools.
- Teams need to rapidly develop shared purpose, vision, trust, and competence to accomplish complex, nonroutine tasks.

As tested and practiced in the complex setting of noncombat JIIM activities and in healthcare, the ToL methodology leads to efficient performance and high-performing leader teams much more quickly than traditional team-building techniques (Prevou et al., 2009). These high-performing teams share certain characteristics:

- They communicate and collaborate more effectively.
- They have a stronger sense of shared purpose, shared trust, shared team competency, and shared team confidence.
- They possess the ability to quickly build strong relationships, empathize with other perspectives, negotiate, and co-opt others (both inside and outside the organization).

Furthermore, ToL combines seven key components: *information technology*; increased *collaboration* as a result of sound knowledge management practices; shared *purpose*, *trust*, *people* (relationships); effective *time* management; and the *links* to resources and information (Figure 15.2). These components increase the reach of the team across the boundaries of geography and time. They facilitate the sharing of critical knowledge required to bring the expertise and experience of multiple agencies toward

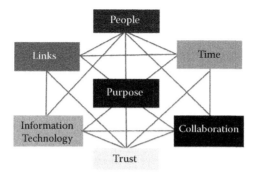

FIGURE 15.2
Components of a *team of leaders* team.

actionable understanding so they may bring the weight of their organizational resources to bear on a problem.

To effectively use all seven components, a problem-solving activity called the *Leader Team Exercise* (LTX) helps the team understand and work through problem resolution and action. This methodology uses a Socratic questioning technique that generates conversation and clarity around both situation and requirements. It helps team members with different perspectives and agendas find common agreement and synchronize efforts. It helps the team collaborate on expected actions and the solutions needed. Finally, it helps the team develop the essential qualities of shared purpose, trust, competence, and confidence required for high performance. The LTX was designed for use both while the team is operating to solve immediate problems and during a preparatory period to build shared qualities before work begins.

KNOWLEDGE MANAGEMENT AND HIGH-PERFORMING TEAMS

At a time when effective multidisciplinary teams are in high demand, the onset of the Information Age has ushered in a period of unparalleled data exchange as well as the need to manage the flow of knowledge through digestible processes. Information technology (IT) itself does not ensure the timely delivery and adequate application of knowledge and information. Without good knowledge management processes to harness and distribute useful information, IT is useless. And, without rapid, experienced decision making, that same useful information is not fully leveraged or is overlooked altogether. Within leader teams operating in a competitive environment, the application of knowledge management is critical, and the storage of information is fleeting, difficult, or arguably irrelevant as was best described by Friedman (2007):

> So you are never out. You are always in. Therefore, you are always on. Bosses, if they are inclined, can collaborate more directly with more of their staff than ever before—no matter who they are or where they are in the hierarchy. But staffers will also have to work much harder to be better informed than their bosses. There are a lot more conversations between bosses and staffers today that start like this, "I know that already! I Googled it myself. Now what do I do about it?" Sort that out. (Friedman, 2007, p. 213)

As an added obstacle to the traditional application of knowledge management, each organization has different capabilities, processes, and restrictions that influence how it shares and collaborates. Within today's flatter, more interconnected world as described by Friedman, the problem before us is a "perfect storm" of sorts between technological incongruity, inconsistent process, and cultural and experiential friction. The mere act of knowing within stovepiped functionality can no longer be the standard to which we strive, particularly in times of crisis. We must continue to ask who else needs to know and who else can and should be contributing. High-performing leader teams must therefore address the areas of IT, KM, and decision making in unison with care and deliberation. We must employ a knowledge system that enables the push, pull, and prodding of key actionable information so that we can decide and act accordingly. The result is a synergistic effect that accelerates the leader team to higher levels of performance. This effect is compounded when crossing more than just organizational boundaries, but exponentially more important when including levels of horizontal and vertical boundaries both within and across organizations as well as move to higher levels of performance.

Knowledge management becomes the bridge that connects the seven components of a ToL team (Figure 15.2) to one another. Referring back to the knowledge environment—people, process, technology, structure, content, culture, and knowledge leadership—framework, an assessment can be made on the lines that connect each component of the knowledge environment to each component of the ToL framework. First within the box itself, then applied box-to-box, and finally expanded across multiple boxes to create synergy. To start applying the framework and understanding who else the team needs to connect to applies the people aspect of the knowledge management framework. From there, building out the connections to important links, how often the underlying processes need to occur will drive the timelines and types of collaboration. From there, understanding whether those connections need to occur physically, virtually, synchronously, or asynchronously, and then the type of technology employed becomes easier to identify.

Furthermore, the level of shared trust between team members and external contributors and the purpose for connecting will drive the timeliness and type of technology employed. Virtual meetings are not nearly as effective in an environment of low trust and may influence the use of technology. For example, a video teleconference that allows face-to-face communication may be essential for a team still establishing shared trust and purpose compared to a team that has already achieved high-performing characteristics and can meet on short notice over a chat or even by e-mail. The key to success is to apply the knowledge management framework against the seven components in order to understand their needs throughout the team's development, starting with the people who make up the team and moving on from there. KM should be centered on the human dimensions of collaboration, communication, sharing, and learning. Knowledge is in your head, affected by attitudes and perceptions from culture and experiences. Therefore, building a strategy needs to take the human dimension into account, and then the process and technology, the organizational structure, content and culture, and knowledge leadership.

For example, if the leader team was made up of team members from different parts of the world, then the components of trust and time may mean different things to various team members. Western society tends to view time as a precious resource that requires the maximum efficiency possible, even at the expense of other components, such as trust. This leads to a heavy reliance on mobile technology and short, time-effective meetings. On the other hand, many cultures around the world view interpersonal relationships as more important than time, necessitating a different style of collaboration and underlying processes. In the end, knowledge management is an essential component to the success of a team because it connects the art and science of making a team function.

RECENT RESEARCH AND PRACTICE ON IMPLEMENTING THE TEAMS OF LEADERS METHODOLOGY

In the past two or more years, we observed teams of leaders application in four major areas with the purpose of learning and adapting the approach:

- *Interagency teams primarily formed within the U.S. government*—A conglomerate of government agencies, including the Departments of State, Agriculture, Justice, and Defense, along with the U.S. Agency for International Development and other, smaller groups.
- *Public health teams oriented on community health improvement*—An all-volunteer, multidisciplinary team brought together because of altruistic feelings or fiscal motivation to improve the community.
- *Military teams, both hierarchical and matrixed*—An Army Brigade Combat Team preparing to deploy.
- *Medical research teams*—Instead of the usual scientists and medical researchers, this group also included a wider array of clinical specialists, therapists, and technicians to round out an Interdisciplinary Team.

The ToL methodology was applied across the different organizations as best made sense within each organization's cultural contexts, but they all collectively employed a similar approach when it came to guided facilitation and coaching to develop shared and actionable understanding in three critical areas required for successful team operations:

- High-performing team qualities, understanding
- Use of information technology systems
- Development of processes and procedures to enable knowledge flow and continuous learning in the organization

In many cases, the formal application of ToL involved employing some form of a *Team Launch Workshop*. This workshop generally consisted of a 2-day event designed to align team members with the seven components illustrated in Figure 15.2. A critical part of the workshop is the *Leader Team Exercise* (LTX), which is a structured thinking process used to facilitate conversation and collaboration. Additionally, for each of the workshops referenced in the case studies, participants and observers completed a 24-question (later 35-question) pre- and postworkshop

Likert scale survey to assess high-performing team attitudes and behaviors. The purpose of the initial survey was to establish a baseline of the team's characteristics. This baseline was then compared to the results of the same survey, administered after the team launch, exercises, and workshop were complete. The key findings from two of the case studies are discussed in more detail later in this chapter. The procedures and results of all four case studies are described more fully in the 2011 paper presented at the Interservice/Interagency Training and Simulation and Education Conference (I/ITSEC) (Prevou et al., 2011).

TEAM DEVELOPMENT LIFE CYCLE

All teams must start somewhere and move toward an objective. Teams of leaders are no different. While they follow a common development process and have a life cycle, the ToL approach facilitates the process so it develops faster and more effectively. The integration of IT and KM further expands the reach and ability to manage critical knowledge, which also facilitates growth, especially across time and space. The LTX acts as an accelerant to team performance by increasing quantity and quality of communication and collaboration, which results in better situational awareness and actionable understanding.

In Figure 15.3, the red line shows normal team development and the blue line shows the desired developmental curve achieved by the ToL approach. Each team must traverse four basic stages: getting started, forming the team, doing the work, and sustaining the team. While most teams may experience a variant of each stage, it is important to remember that merely traversing the four stages does not propel a team to high performance. If it does achieve a level of high performance, it is usually a delayed effect that happens well into mission execution. The LTX drives steep curve increases throughout the team's life cycle.

Every team will progress through the four stages at its own pace and develop its own culture, and the groundwork laid by the LTX methodology provides the mechanics to constructively address challenges and build positive relationships. Through each stage, understanding, purpose, and wisdom are passed throughout the team. In this context, IT and KM become enablers of the LTX by providing the team the means to connect and manage high volumes of information.

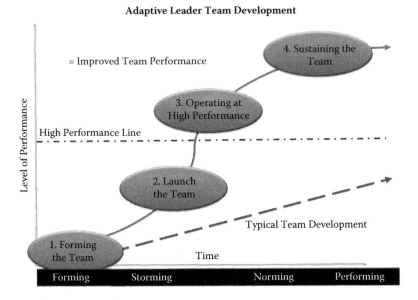

FIGURE 15.3
Team development life cycle.

Stage 1: Getting Started

Stage one consists of those tasks done to prepare the leader team. After identifying the initial mission and the specific team members, it is imperative that the leader team's organizer make initial contact as soon as possible with potential members of the team. It is critical to lay the groundwork for how business will be conducted.

Stage 2: Forming the Team

Stage two is arguably the most important stage, for it is here that the team of leaders develops the basic understandings and agreements necessary for success. Leader team members begin forming a shared understanding of the mission. Individual skill sets are identified as well as those efforts required to develop shared skills. As Hackman (2002) noted, most organizations do not know how to launch a team, and this is a significant doctrinal shortcoming.

Stage two is very much a sense-making and consensus-building stage. Operating agreements are formed and the team begins to understand specific boundaries. It will be the first time the team is faced with challenges and disagreements. Working relationships begin to emerge. The impact of stage two will likely define the long-term environment in which the team will operate. "If we fail to get stage 2 about right, then we could be expending a lot of energy in the wrong direction" (Hilton, 2009a). If the groundwork from stage two has not been laid, the team stands to charge headlong into its mission without a clear direction.

Stage 3: Doing the Work

It should be noted that leader teams may be required to move through this stage rapidly depending on the environment they are operating within and the pressure to provide deliverables or reach objectives. Positive working relationships, communication and collaborative processes, and operating agreements are all established as the team moves into the execution phase. The LTX provides a nonconfrontational framework to address major challenges and to build understanding, if not consensus. As operational challenges arise, the team may revisit Stage 2 and modify or adjust as the team becomes aware of previous unknowns. In this stage we observed the greatest improvements in trust and team confidence.

Stage 4: Sustaining the Team

Unfortunately, most established teams overlook this critical stage. Once a team achieves high performance, the tendency is to take a "breather." Yet, many events or activities can reduce the team's impact over time. Changes in the situation or mission, fluctuating membership, and the absorption or reduction of technology and processes may seem inconsequential but can result in long hours of confusion and course correction. In this stage, the LTX helps the team deal with each of the situations above. The team should never hesitate to drop back to a previous stage to reaffirm its procedures and agreements. Our observations were that teams that do go back to recalibrate, rebound quickly, while those that do not begin a downward curve.

THE LEADER TEAM EXERCISE

At the heart of the teams of leaders approach is the Leader Team Exercise (LTX). Based on Gary Klein's model for building intuitive decision making (2003), the exercise provides the leader team with the tool it needs to punch through most barriers and boundaries. Rather than the

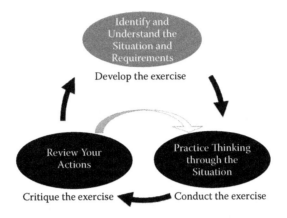

FIGURE 15.4
The leader team exercise (LTX). (Adapted from Klein, G.A., *The Power of Intuition*, Currency Books, NY, 2003.)

traditional "learn, train, do" cycle that pauses productivity, the LTX is conducted in real time as part of the operational decision-making process and draws out the team's holistic experience and ability. It is a way of understanding and working through problem sets and decisions as a group, rather than clear chains of command. Leader teams may agree on the desired outcome, but they often have philosophical differences with respect to approach. The LTX helps build a shared mental model that focuses the teams' vision, effort, and ability with reduced friction so they may come to an agreement faster. In the process, it builds a shared trust by drawing out the collective competence of the team resulting in the confidence the team needs to address any situation. Ideally, this process (shown in Figure 15.4) would be coached until it is second nature to the team.

Step 1: Identify and Understand the Situation and Requirements

This step asks the team, as a group, to simply describe the situation as they believe it exists and what it is they are trying to accomplish. Before any team can act, they must agree on the problem and the desired end state. Often, teams are thrust together with little time to react or prepare. The natural instinct is to dogmatically approach the problem from one's own experience. Problems arise when teams fail to understand exactly what the situation requires. As members perform this step, they begin building a common appreciation of other members' intuition and the foundation is laid for a common vision or purpose. The results of this step are a common understanding of where the team members agree and disagree about situation and requirements.

Step 2: Practice Thinking through the Situation

In this step, members discuss how they visualize the situation unfolding and what must be done to accomplish the requirements identified in Step 1. The team then conducts "what-if" drills that force each member to form mental models that may be outside of their realms of experience. This step increases understanding of each organization's capabilities and improves adaptability and problem solving. The outcome of this step is a common understanding of what might happen as the situation plays out and what each organization represented by the team members can offer to its successful resolution.

Step 3: Review Your Actions—Adjust as Needed

Here the collective knowledge is harnessed as each member contributes his or her unique gift to the solution. After conducting "what-if" drills, the team conducts "what-then" drills. They begin to understand the potential pitfalls before them. As those pitfalls are identified, the cycle repeats itself until all "what-if," "what-then" scenarios are satisfied. Team members codify who will do what, changes to operating agreements, and points of friction.

Building shared skills (behaviors), knowledge, and attitudes drives high levels of adaptive leader-team performance. Such leader-team development requires experiential learning based on iterative, coached, and deliberate practice found in the ToL approach LTX. This exercise provides a structured yet flexible way to think through the situation, requirements, decisions, and responsibilities in a matter of minutes, while exposing different agendas, views, and capabilities. Because the LTX is accomplished while the team is working, it facilitates learning while doing and is easily incorporated into today's high operating tempo.

CASE STUDIES

Applying the Teams of Leaders approach has appeared beneficial across multiple diversified organizations. The following two case studies demonstrate examples of ToL and the LTX action applied by each organization within their unique cultural and organizational dynamics. The cases highlight the teams of leaders approach across interagency teams within the whole of government and the military, and cross-function and cross-county teams in Michigan to help deal with the complexities of public health. Employing ToL should not be limited to these two approaches because flexibility to account for the endless number of variables within the human dimension is perhaps one of the approach's greatest strengths.

CASE STUDY 1   The Military and The Interagency Team

European Command: As discussed earlier there was a ToL initiative ongoing at the U.S. European Command (EUCOM). During an annual exercise called AUSTERE CHALLENGE, which took place in Germany in April and May 2009, ToL was employed to foster teaming among a conglomerate of government agencies, including the Departments of State, Agriculture, Justice, and Defense, along with the U.S. Agency for International Development and many other, smaller groups.

During the exercise, the research team observed and evaluated the performance of five teams and their interactions, both internally among members, and externally among other leader teams. The main focus was on two interagency teams deployed from Washington, DC, and three EUCOM Headquarters internal teams. Each leader-team was composed of leaders representing their respective organization or agency from multiple levels of government. The results of this study were published at IITSEC in 2009 (Prevou et al., 2009).

The leader teams were measured for effectiveness through direct observation and scored on three different occasions using six criteria and informal interviews. The research team used a 24-question assessment from the ToL coaching guide that measured shared purpose, trust, competence, confidence, collaborative technology, information, and knowledge management. (This instrument has since been updated.)

Of the five teams, Leader Teams 1 and 2 had neither ToL training nor coaching, while ToL interventions were performed on Leader Teams 3, 4, and 5 (Figure 15.5).

Team Performance Graphic - Case 1

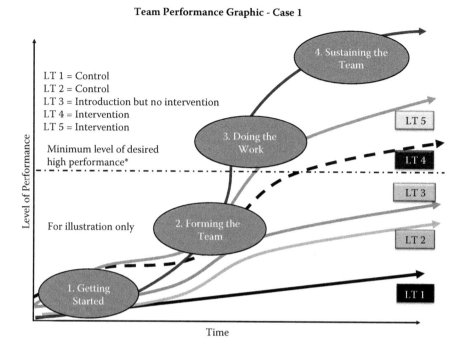

FIGURE 15.5
Test group performance.

Leader Teams 1 and 2, with no ToL training, served as the control groups. They formed and operated as most teams do with no formal launch, operating procedures, or coaching, struggling throughout the 10-day exercise to reach even modest levels of performance.

Leader Team 3 received a briefing and short, 2-hour class on the ToL approach. Following observation, they were noted by the research team as having some of the best collaborative processes; however, the team's organization and synchronicity were initially lacking. At the outset, a disagreement on purpose and a problem of trust with other teams was observed. Left unresolved, these problems hampered the team's overall performance.

Leader Team 4 received the same briefing and short class on ToL, but was also assigned a ToL coach from outside the organization as group facilitator. This additional coaching resulted in a solid level of sustained performance. The team demonstrated shared understanding of the situation, purpose, and trust. They adequately achieved the desired end state, but they experienced some difficulty in effectively using collaborative technologies. These difficulties were somewhat mitigated by the constructive relationships between the key members, team experience, and coaching in the ToL methodology.

Leader Team 5 received the briefing and ToL class but had an internal (rather than external) coach assigned as group facilitator. This coach previously completed a day of ToL training, making him comparable in skill with the external coach assigned to Team 4. Through coaching and the LTX, this team was able to achieve the highest levels of performance. Virtual collaboration, use of the team information portal, and knowledge management among the members were some of the best observed.

Figure 15.5 compares the performance of each team during the exercise against the dashed horizontal line for "minimum level of performance" and the red "S" curve line, which represents desired performance. The blue ovals on the red "S" line indicate the desired stages of a ToL team.

From Figure 15.5, it is clear that each team that applied even the smallest parts of the ToL approach fared better than the teams who lacked ToL knowledge. The teams that applied the ToL approach haphazardly realized modest improvements over 10 days but failed to reach the standard for high team performance set by unit leaders. Teams that applied ToL with the assistance of an expert facilitator showed the most improvement.

CASE STUDY 2   Public Health Teams in Central Michigan

Early in 2010, a series of reports were released by the University of Wisconsin's Population Health Institute, which documented the overall "health" of counties across the United States. In the report, six counties under the purview of the Central Michigan Health District performed poorly and spurred the District Health Officer to action, convening a public health summit assembling health providers, social workers, mental health professionals, leaders from business and government, and the media to discuss strategies to improve overall public health. The group came together in January 2011 for a Team Launch Workshop and to develop a core group of internal ToL coaches/trainers who would stay within each county to facilitate future activities.

This was a multidisciplinary group, and all team members were volunteers—there because of altruistic feelings or fiscal motivation to improve the community. Although the initial meetings were considered ineffective, the group shared motivation, an understanding of purpose, and the mission of "Uniting the communities and working together, we will improve health and promote wellness in central Michigan." This mission statement, a vision ("Together We Can Build a Healthier Community"), and core values were developed during the workshop along with a set of formal operating agreements. These documents—the vision, mission, values, and operating agreements—formed the basis of an orientation package available to on-board new team members. The package was also used as a basic team charter as the group moved the process into each of the individual counties.

During the workshop, the team developed a set of eight short-term and long-term objectives to assist them in working with the six counties, as well as developing a district-wide health improvement plan. Comments at the end of the workshop indicated the enthusiasm and momentum generated.

Slightly modified versions of LTX and the pre- and post-assessments were given to the team members (purpose, people, links, time, and trust were assessed), and most measures showed improvement.

Information technology and knowledge management were not directly measured, but they played a vital role in group discussion. During discussion, it became evident that the team—like most multidisciplinary teams—had no common platform for information exchange, and no formalized knowledge management processes in place. It was determined that these two issues would greatly hamper future cross-boundary communication and should be remedied. We are currently working with this group to establish a collaborative communications platform (or Team Room) where members can continue to work together, share information, and work collaboratively on developing the health improvement plan and share resources across the counties.

Lessons from use in Michigan:

- Team charters, which codify vision, mission, and values, as well as operating agreements are critical to the success of disparate teams with frequently changing membership.

- Lack of common team platforms, methods, and protocols for information exchange hamper team performance even when other team performance measures are high.

The LTX is an effective and powerful tool to develop shared understanding of purpose and build trust through a structured and deliberate conversation. Even where the team believed they had an understanding of the overall vision and mission at the beginning of the workshop, there was still positive movement along the dimensions of purpose on the postassessment.

Both the LTX and deliberate operating agreements are especially important drivers of shared trust and confidence for disparate teams. The rich conversation produced by the LTX, combined with discussion and agreement about the team's "rules of engagement," helped the team members build relationships with one another.

FUTURE RESEARCH NEEDED IN THE AREA

Although initial observations on teams of leaders implementation generated positive results, more work is needed to fully understand the breadth and depth of the benefits and disadvantages of the methodology. The teams of leaders approach is inexpensive when compared to technology-based solutions or the cost of a poorly performing team. The leader team exercise also appears beneficial as a way to rapidly and effectively work through any type of problem, generating understanding and trust in a true collaborative spirit. However, the exercise may require some level of upfront coaching and facilitation to learn effective employment techniques. Because ToL is relatively young, the lack of longitudinal data does not show the long-term impacts on repeated leader team formation and if the associated barriers to high performance lessen over time as it integrates into the fabric of the organization's culture. At the present time, we believe the approach must be reinforced by a ToL coach until the team has mastered the basics. The ToL approach leverages technology and knowledge management to improve team outcomes. This, too, deserves some further research—not just on how we team, but the relationship in employing leader teams in an environment increasingly influenced by Generation Y and the maturation of the social networking movement. Finally, perhaps the biggest challenge for future research is the cultural impacts within an organization to unleash the inherent power of social networks within agreed upon rules of engagement to allow the formation of informal teams with a robust organizational support structure in place. We envision an organizational culture where teams can form rapidly and naturally, have the resources (IT and KM) always available, and have a cadre of leaders who can effectively coach a new team through the form and launch stages of their life cycle.

CONCLUSION

Because we are under increasing pressure to evolve to become more responsive to the rapidly changing rate of global interconnectivity, we need to rethink how our organizations form and launch very diverse cross-boundary teams. We need to open our minds about what is required to provide organizational support and coaching to these teams and how we will facilitate connecting the team in a safe and secure way. The viral development and pervasive integration of social networking changed the level of involvement and access to knowledge, which is unparalleled to any other time in our world's history. That connectivity exposes new risks to rapidly elevate local or regional issues quickly onto the global stage and challenges society's traditions and cultural norms, while at the same time offering opportunity to those daring enough to take advantage of changing times for competitive advantage. It also increases the speed of competition and market decision so that old, tried-and-true methodologies may not be sufficient to keep us competitive.

In response, the traditional knowledge management framework needs to be reevaluated and reoriented on the human dimension and the centrality of people within a broader and more comprehensive knowledge environment framework. More technology will not solve challenges in connecting people to one another, especially when crossing traditional boundaries of culture and organization. To account for these changes, expansion is needed within the knowledge environment, which includes the dynamics of organizational culture, structure, and knowledge leadership as necessary to remain relevant in a networked, flattened world. Today as well as tomorrow, organizations must operate across time and space, effectively connecting one another to collaborate more effectively and build relationships and understanding toward solving complex problems and facing tough issues in the future.

Organizations seeking to leverage the power of knowledge management and social networking need to look no further than the leader teams that exist within their organization. The potential of a team's performance in the corporate, academic, and military world is too important to leave to happenstance. The need for shared understanding and vision, well-defined operating procedures (business rules), team responsibilities, and the processes for using the IT/KM tools, such as knowledge portals, Web conferencing, content management, wikis, and shared-rolodexes to collaborate and communicate, cannot be understated. Most importantly, the ability to form enduring, flexible relationships across cultural and regulatory boundaries is paramount to collaborative team success.

We do not practice KM for the sake of KM. It must be linked to specific outcomes and help an organization, team, or individual be more competitive. Any deliberate approach, such as teams of leaders, that improves the way we form and launch teams, communicate, and collaborate will offer opportunities to grow and remain competitive and relevant.

REFERENCES

Bradford, Z.B., Jr., and Brown, F.J. (2007). Teams of Leaders: The Next Multiplier. *Landpower Essay 07-1*. Arlington, VA: Association of the U.S. Army.

Bradford, Z.B., and Brown, F.J. (2008). *America's Army, A Model for Interagency Effectiveness*. Westport, CT: Prager Security International.

Brown, F.J. (2009). Teams of Leaders in U.S. European Command: A Soft-Power Multiplier. *Landpower Essay 09-2*. Arlington, VA: Association of the U.S. Army.

Coutu, D. (2009). Why teams don't work. An interview with Richard Hackman. *Harvard Business Review,* May, 99–105, 131.

Dempsey, M.E. (2009, May 13). *Joint Warfighting Development*. Joint Warfighting Conference. Podcast retrieved from http://www.army.mil/-news/2009/05/12/21007-developing-leaders-is-job-1-at-tradoc/.

Friedman, T.L. (2007). *The World Is Flat [Updated and Expanded]: A Brief History of the Twenty-First Century*. New York: Farrar, Straus and Giroux.

Hackman, J.R. (2002). *Leading Teams: Setting the Stage for Great Performance*. Boston, MA: HBSP.

Hilton, B. (2009a, January). Conversation at coordination meetings at EUCOM HQ, Stuttgart, Germany.

Hilton, B. (2009b, January). Conversation at a EUCOM meeting on Teams of Leader implementation.

Hilton, B. (2009c). Enabling Collaboration through Teams of Leaders—ToL. In O'Connor, J., Bienenstock, E.J., Briggs, R.O., Dodd, C., Hunt, C., Kiernan, K., et al. (Eds.), Collaboration in the National Security Arena: Myths and Reality—What Science and

Experience Can Contribute to Its Success (pp. 33–39). Washington, DC: Strategic Multi-Layer Assessment (SMA) effort, synchronized by Joint Staff executed through STRATCOM/GISC and OSD/DDRE/RRTO.

Klein, G.A. (2003). *The Power of Intuition.* New York: Currency Books.

Leistner, F. (2010). *Mastering Organizational Knowledge Flow: How to Make Knowledge Sharing Work.* New York: SAS Institute and Wiley.

Lipnack, J., and Stamps, J. (2000). *Virtual Teams: People Working Across Boundaries with Technology,* 2nd ed. New York: Wiley.

Neissen, M. (2006). *Harnessing Knowledge Dynamics: Principled Organizational Knowing and Learning.* Hershey, PA: IRM Press.

Prevou, M. (2011). Understanding the Knowledge Environment. *Army Communicator, Summer 2011.* Fort Gordon, GA: U.S. Army Signal Center. http://www.signal.army. mil/ArmyCommunicator/2011/Vol36/No2/2011Vol36No2Sub17.pdf

Prevou, M., Hilton, B., Hower, M., McGurn, L., and Gibson, C., (2011). Building effective teams faster. Paper no. 11242. *Proceedings of the Interservice/Interagency Training and Simulation and Education Conference (I/ITSEC),* Orlando, FL.

Prevou, M., Veitch, R., and Sullivan, R. (2009). Teams of leaders: Raising the level of collaborative leader-team performance. Paper no. 9089. Proceedings of the Interservice/ Interagency Training and Simulation and Education Conference (I/ITSEC), Orlando, FL.

Senge, P. (1990). *The Fifth Discipline: The Art and Practice of the Learning Organization.* New York: Doubleday.

U.S. Army. (2008, August). *FM 6.01.1. The Knowledge Management Section.* Ft. Leavenworth, KS: U.S. Army Combined Arms Center.

U.S. Army. (2009, June). *Leader Development Strategy, Draft V13 (U.S. Army LDS).* Ft. Leavenworth, KS: Center for Army Leadership and Command and General Staff College, U.S. Army Combined Arms Center.

U.S. Army. *Mission Command Overview Briefing Slides.* Ft. Leavenworth, KS: Mission Command Center of Excellence, U.S. Army Combined Arms Center.

U.S. Army. (2009, June). Teams of Leaders Handbook. vol I. Ft. Leaven worth KS: Battle Command Knowledge Systems (BCKS), U.S. Army Combined Arms Center.

Willyerd, K., and Meister, J.C. (2009, November 11). The Über-Connected Organization: A Mandate for 2010. Retrieved from HBR Blog Network, http://blogs.hbr.org/ cs/2009/11/the_uberconnected_organization.html.

The White House Archives. (2006). The Federal Response to Hurricane Katrina Lessons Learned, Chapter Six: Transforming National Preparedness. http://georgewbush-whitehouse.archives.gov/reports/katrina-lessons-learned/chapter6.html.

Index